The Role of Biofilms in the Development and Dissemination of Microbial Resistance within the Food Industry

The Role of Biofilms in the Development and Dissemination of Microbial Resistance within the Food Industry

Editors

Efstathios Giaouris
Manuel Simões
Florence Dubois-Brissonnet

MDPI • Basel • Beijing • Wuhan • Barcelona • Belgrade • Manchester • Tokyo • Cluj • Tianjin

Editors
Efstathios Giaouris
University of the Aegean
Greece

Manuel Simões
University of Porto
Portugal

Florence Dubois-Brissonnet
AgroParisTech
France

Editorial Office
MDPI
St. Alban-Anlage 66
4052 Basel, Switzerland

This is a reprint of articles from the Special Issue published online in the open access journal *Foods* (ISSN 2304-8158) (available at: https://www.mdpi.com/journal/foods/special_issues/Food_Industry).

For citation purposes, cite each article independently as indicated on the article page online and as indicated below:

LastName, A.A.; LastName, B.B.; LastName, C.C. Article Title. *Journal Name* **Year**, *Article Number*, Page Range.

ISBN 978-3-03943-551-7 (Hbk)
ISBN 978-3-03943-552-4 (PDF)

Contents

About the Editors

Efstathios Giaouris is currently serving as Associate Professor in Food Microbiology with the Department of Food Science and Nutrition of the University of the Aegean (Lemnos, Greece). His main research interests are focused on pathogenic bacterial biofilms, and especially the risk of this inherent microbial lifestyle for food hygiene and safety, and their control using novel, cost-efficient, and sustainable methods (e.g., phytochemicals, enzymes, QS inhibitors and nanomaterials). He is reviewer and specialist expert of research proposals to national and international organizations and registered Trainer of Hellenic Food Authority (EFET) on topics of food hygiene and safety, of staff working in the food industry and food control agencies. He has published 37 articles in international scientific peer-reviewed Journals, 7 chapters in collective English volumes, and he has also announced his research projects in several national (Greek) and international conferences (more than 20 orals and more than 50 poster presentations) (h-index = 17).

Manuel Simões graduated in Biological Engineering and received a Ph.D. in Chemical and Biological Engineering. Currently, he is an Assistant Professor, Pro-Director of the Faculty of Engineering at the University of Porto (FEUP), country, and member of the Laboratory for Process Engineering, Environment, Biotechnology, and Energy (LEPABE) in the Department of Chemical Engineering at FEUP. Manuel Simões has published more than 160 papers in journals indexed in JCR (h-index = 42), 4 books (1 as author and 3 as editor), and more than 40 book chapters. He is an Assistant Editor of *Biofouling–The Journal of Bioadhesion and Biofilm Research* (the oldest journal on biofilm research) and Associate Editor of *Frontiers in Microbiology*. His main research interests are focused on the mechanisms of biofilm formation and their control with antimicrobial agents, particularly using new antimicrobial molecules.

Florence Dubois-Brissonnet is Professor of Microbiology and Food Safety with the Department of Food Science and Technology of AgroParisTech and at Micalis Institute (INRAE) in Jouy-en-Josas (France). She coordinates research programs on adaptation and persistence of foodborne pathogens in food processing environments. The membrane physiology is an indicator of particular interest that she uses to better understand the adaptation of food-borne pathogens in foods and biofilms and their resistance to disinfectants and natural antimicrobial compounds. She has completed more than 60 publications or book chapters and provided more than 85 oral or poster presentations in congress (h-index = 19). She has coordinated a book published in 2017 in Lavoisier collection about microbiological food risk in French, "Risques microbiologiques alimentaires". She is vice-president of the European ISEKI-Food Association, which networks universities in food studies.

Preface to "The Role of Biofilms in the Development and Dissemination of Microbial Resistance within the Food Industry"

Biofilms are multicellular sessile microbial communities embedded in hydrated extracellular polymeric matrices. Their formation is common in microbial life in most environments, while those formed on food-processing surfaces are of considerable interest in the context of food hygiene. Biofilm cells express properties that are distinct from planktonic ones, in particular, notorious resistance to antimicrobial agents. Thus, a special feature of biofilms is that, once they have been developed, they are hard to eradicate, even when careful sanitization procedures are regularly applied. A great deal of ongoing research has investigated how and why surface-attached microbial communities develop such resistance, and several mechanisms are to be acknowledged (e.g., heterogeneous metabolic activity, cell adaptive responses, diffusion limitations, genetic and functional diversification, and microbial interactions). The articles contained in this Special Issue deal with biofilms of some important food-related bacteria (including common pathogens such as *Salmonella enterica*, *Listeria monocytogenes*, and *Staphylococcus aureus*, as well as spoilage-causing spore-forming bacilli), providing novel insights on their resistance mechanisms and implications, together with novel methods (e.g., use of protective biofilms formed by beneficial bacteria, enzymes) that could be used to overcome such resistance and thus improve the safety of our food supply and protect public health.

Efstathios Giaouris, Manuel Simões, Florence Dubois-Brissonnet
Editors

Editorial

The Role of Biofilms in the Development and Dissemination of Microbial Resistance within the Food Industry

Efstathios Giaouris [1],*, Manuel Simões [2] and Florence Dubois-Brissonnet [3]

[1] Laboratory of Biology, Microbiology and Biotechnology of Foods (LBMBF), Department of Food Science and Nutrition, School of the Environment, University of the Aegean, Ierou Lochou 10 & Makrygianni, 81400 Myrina, Lemnos, Greece

[2] Laboratory for Process Engineering, Environment, Biotechnology and Energy (LEPABE), Department of Chemical Engineering, Faculty of Engineering, University of Porto, Rua Roberto Frias, s/n, 4200-465 Porto, Portugal; mvs@fe.up.pt

[3] Micalis Institute, INRAE, AgroParisTech, Université Paris-Saclay, 78350 Jouy-en-Josas, France; florence.dubois-brissonnet@agroparistech.fr

* Correspondence: stagiaouris@aegean.gr; Tel.: +30-22540-83115

Received: 15 June 2020; Accepted: 16 June 2020; Published: 21 June 2020

Abstract: Biofilms are multicellular sessile microbial communities embedded in hydrated extracellular polymeric matrices. Their formation is common in microbial life in most environments, while those formed on food-processing surfaces are of considerable interest in the context of food hygiene. Biofilm cells express properties that are distinct from planktonic ones, in particular, notorious resistance to antimicrobial agents. Thus, a special feature of biofilms is that, once they have been developed, they are hard to eradicate, even when careful sanitization procedures are regularly applied. A great deal of ongoing research has investigated how and why surface-attached microbial communities develop such resistance, and several mechanisms are to be acknowledged (e.g., heterogeneous metabolic activity, cell adaptive responses, diffusion limitations, genetic and functional diversification, and microbial interactions). The articles contained in this Special Issue deal with biofilms of some important food-related bacteria (including common pathogens such as *Salmonella enterica*, *Listeria monocytogenes*, and *Staphylococcus aureus*, as well as spoilage-causing spore-forming bacilli), providing novel insights on their resistance mechanisms and implications, together with novel methods (e.g., use of protective biofilms formed by beneficial bacteria, enzymes) that could be used to overcome such resistance and thus improve the safety of our food supply and protect public health.

Keywords: biofilms; foodborne pathogens; dairy bacilli; stress adaptation; resistance; disinfection; biocontrol; enzymes; food safety

---

The formation of biofilms spontaneously happens in both natural and industrial environments, wherever there are microorganisms, surfaces, nutrients, and water. In previous years, many studies have been occupied with detrimental biofilms, such as those formed by/containing pathogenic microorganisms, providing enough useful data on the complex mechanisms that may account for their increased recalcitrance towards antimicrobials, host immune system, and many other physicochemical stresses. Thus, diffusion limitations to the free access of some antimicrobials inside the robust biofilm matrix, variability in the physicochemical microenvironments within the biofilm (e.g., pH, oxygen levels, nutrients), cellular adaptations resulting from altered gene expression and/or horizontal gene transfer, microbial interactions, and the differentiation of biofilm-enclosed microbial cells into particularly durable variants, such as viable but not culturable (VBNC) ones, and persisters, may all account, at different levels and depending on the specific microorganisms and the surroundings, to the robustness of biofilms [1]. Their establishment as the default mode of microbial growth is hence almost everywhere. Biofilms formed by pathogenic bacteria are of special interest in the context of food hygiene since they may significantly compromise food safety [2]. Those containing spoilage microflora can downgrade food quality, limiting shelf life of the products, and induce several other important issues (e.g., clogging of membranes, increases in energy costs, biofouling, and corrosion problems). In the articles of this Special Issue, interesting data are presented regarding such biofilm communities towards the better understanding of the factors that can influence their sessile development (e.g., microbial interactions, sporulation, food residues, temperature), the mechanisms lying behind their antimicrobial resistance, together with some novel alternative methods that could be exploited to address this important problem (e.g., use of lactic acid bacteria and/or their metabolites, enzymes, bacteriophages, quorum sensing inhibitors), with lower possibilities for resistance occurrence.

Bacillus species are frequently encountered in the dairy processing environment and can form biofilms on surfaces containing their spores, and in this way, resist cleaning-in-place (CIP) regimes commonly applied in the dairy industry. Those consist of regular cleaning of equipment with alkaline and acidic solutions under turbulent flow conditions at high temperatures. Ostrov et al. [3] investigated the resistance of biofilm-derived spores of four dairy-associated *Bacillus* isolates (including one *B. licheniformis*, one *B. subtilis*, and two *B. paralicheniformis* strains) to CIP procedures and compared to those of a non-dairy *B. subtilis* isolate, using in parallel two different model systems simulating the typical conditions for milking systems. As cleaning solutions, they used caustic soda (0.5% w/v NaOH), sodium hypochlorite (0.018% v/v NaOCl), and six different commercial alkaline detergents commonly used in dairy farms and at concentrations recommended by the manufacturers. They observed that the dairy-associated isolates displayed increased resistance to mechanical (i.e., water circulation), chemo-biological (i.e., cleaning), and bactericidal (i.e., disinfection) effects of the tested CIP procedures compared to the non-dairy *Bacillus*. This was attributed to their robust biofilm formation and to differences in the structure and composition of their biofilm matrix resulting in its mucoid appearance. This finding was further reinforced by the enhanced resistance of two other poly-γ-glutamic acid (PGA)-overproducing *B. subtilis* strains to the tested CIP procedures, compared to the wild type strain. These mutant strains could indeed produce high amounts of proteinaceous extracellular matrix, which was similar in appearance to that produced by the tested dairy *Bacillus* isolates. The authors highlighted the importance of using strong biofilm-formers, such as biofilm-derived spores of dairy-associated *Bacillus*, upon evaluating the performance of commercial cleaning agents for use in industrial conditions. Undoubtedly, their results seem important towards the refinement of the industrial CIP processes to increase their efficiency in eliminating well established biofilms.

Bovine mastitis is among the most common diseases that the dairy industry should deal with, resulting in considerable economic losses due to milk wastage and treatment costs. This is frequently caused by pathogenic staphylococci capable of forming biofilms inside the udder and making this ineffective the subsequent antibiotic therapy. Wallis et al. [4] evaluated the in vitro efficiency of an alternative therapeutic approach based on the formation of beneficial (probiotic) biofilms by lactic acid bacteria (LAB). For this, they employed five LAB strains (including three *Lactobacillus plantarum*, one *L. brevis*, and one *L. rhamnosus*) and tested them for their ability to eradicate and replace harmful *Staphylococcus* biofilms, formed by three different species all known to be implicated in bovine mastitis (i.e., *Staphylococcus aureus*, *S. xylosus*, and *S. epidermidis*). To do this, they left staphylococci to form biofilms on the wells of polypropylene 96-well plates at 37 °C for 168 h before the addition of each LAB culture and further incubation at 37 °C for 168 h. They removed biofilm cells from surfaces at three different time intervals and enumerated them. They found that all the tested LAB strains were able to remove the pathogenic biofilms, while two of them (*L. rhamnosus* ATCC 7469 and *L. plantarum* 2/37) could also form their own biofilms in the place of the pathogenic ones. The authors concluded that these two LAB strains could be suitable for a probiotic treatment of mastitis, and proposed them for further in vivo investigations to test their potential beneficial/barrier properties on udder health.

The biofilm matrix largely accounts for the reduced efficiency of antimicrobials against the biofilm-enclosed microorganisms by delaying their diffusion, scavenging or even inactivating them, and in parallel altering the local microenvironment of the cells, resulting in their slower growth rate and stress adaptation. This is usually composed of polysaccharides, proteins, lipids, and nucleic acids. Concerning the latter, the presence of extracellular DNA (e-DNA) has been recently been reported as a substantial component of the biofilm matrices of several microorganisms. Since the matrix plays a major role in biofilm stability, keeping it close together and hydrating the microbial cells, its degradation could consist in an effective antibiofilm strategy. This could be achieved by using enzymes targeting its main components. To this direction, Sharma and Pagedar Singh [5] tested the efficiency of DNase against mono- and mixed-species biofilms of some microorganisms relevant to the food industry (i.e., *S. aureus*, *Klebsiella* spp., *Enterococcus faecalis*, and *Salmonella* Typhimurium). First, they optimized the enzymatic treatment against biofilms formed by *Pseudomonas aeruginosa* PAO1, which was used as bacterial model due to its ability to produce copious biofilm. They applied the enzyme during biofilm formation (pre-treatment), following biofilm formation (post-treatment), and both before and after (dual treatment). Pre-treatment of DNase at a concentration of 10 µg/mL reduced biofilm formation by *P. aeruginosa* at 37 °C for 24 h by 70%, with no further efficiency to be observed upon increasing the concentration of the enzyme. Interestingly, DNase was less efficient when biofilms were older (up to 96 h), indicating that mature biofilms are more resistant than those of lower age. Post-treatment for 15 min with the same concentration of the enzyme was proven to be more efficient, resulting in a 73–77% reduction in biofilm biomass, depending on the age of the biofilm (24–96 h). The concomitant presence of Mg^{2+} ions (10 mM), used as cofactors for the enzyme, resulted in 90% reduction of *P. aeruginosa* biofilm at a half concentration (i.e., 5 µg/mL) and irrespectively of the age of biofilm. No significant differences were observed between the pre-, post-, and dual-treatments on mono-species biofilms of all the other bacteria, with their susceptibility to DNase still being organism specific. In addition, DNase was less efficient against 24 h-old mixed-species biofilms compared to mono-species ones, and its efficiency was further reduced when biofilms were grown for 48 h. The authors concluded that further optimization is required before applying DNase in cleaning regimes in food industries targeting both biofilm prevention and reduction of mixed-species sessile consortia.

Salmonella enterica is a major foodborne pathogen, worldwide, being frequently implicated in large outbreaks. Many studies have explored its ability to produce biofilm on either abiotic or biotic surfaces and, like with other microorganisms, this is considered as an important stress adaptation strategy [6]. Paz-Méndez et al. [7] investigated the ability of 13 strains of this pathogen, isolated from poultry houses and belonging to three different subspecies (i.e., *enterica*, *arizonae*, and *salamae*) and nine different serovars (including Typhimurium, Enteritidis, Newport, Infantis etc.) to produce biofilm on two different surfaces (i.e., stainless steel and polystyrene), incubated for 48 h in four different growth media at two temperatures (i.e., 6 °C and 22 °C). The colony morphotypes of these strains and their motilities were also investigated at both temperatures. They found that the diluted laboratory growth medium favored biofilm formation, irrespective of the surface and temperature tested compared to the other media containing food residues and used to simulate growth conditions encountered in the different food industries (i.e., dairy, meat and vegetables). Nevertheless, most of the strains were still able to produce biofilm in the presence of food residues under all the tested conditions. Almost all strains (except two) produced the red, dry, and rough (RDAR) morphotype at 22 °C, whereas a soft and completely white (SACW) morphotype was apparent at the lower temperature (i.e., 6 °C). RDAR morphotype is known to arise due to the production of cellulose and curli fimbriae, which have been both described as the main extracellular polymeric substances (EPSs) of the *Salmonella* biofilm matrix. Indeed, biofilm formation was higher at 22 °C compared to 6 °C, with the exception of tomato juice, where the biomass differences were not significant. However, the fact that most of the strains were still able to produce biofilm at 6 °C implies that other components and genetic mechanisms should play a role in the transition of cells to this sessile lifestyle. Similar to the biofilm-forming capacity, the mean motility of the strains was significantly higher at 22 °C than at 6 °C. The authors conclude that *Salmonella* bacteria may use food residues to produce biofilms on common surfaces of the food chain. Further studies combining more strains and food residues should increase our knowledge on *Salmonella* biofilm behaviour in the presence of such nutrient's sources. These are considered important since they better mimic food industry conditions, which may well differ from those encountered in the laboratory, inducing drastic implications on biofilm/cellular physiology and resistance.

The resistance of *Salmonella* being confined in biofilm structures to disinfectants commonly used during poultry processing is surely an alarming public health issue. The review of Cadena et al. [8] examines the modes of action of various types of disinfectants (including hexadecylpyridinium chloride, peracetic acid, sodium hypochlorite, and trisodium phosphate) against *Salmonella* in either planktonic or biofilm state, and in parallel describes the mechanisms that may confer tolerance to such disinfectants and cross-protection to antibiotics. The authors conclude that poultry processors should try to use various disinfectants presenting different modes of action to limit the ability of the bacteria to adapt and display antimicrobial resistance (AMR). The use of alternative approaches, such as enzymes, bacteriophages, and quorum sensing inhibitors, may also be valuable towards the control of biofilms and food safety assurance with lower probabilities of AMR induction. In addition, since the in situ detection of biofilms is important to be able to optimize the prevention and control methods, some commercially available devices and kits that could be used for either qualitative or quantitative, direct or indirect characterization of biofilms encountered in food processing environments are reported.

This Special Issue finishes with an interesting review presenting an update on our knowledge related to *Listeria monocytogenes* biofilms in food-related environments and their implications mainly towards biocide resistance [9]. Legislation, important ecological aspects (i.e., influence of microbial interactions on resistance in mixed-species biofilms), and some potential biocontrol strategies (i.e., use of lactic acid bacteria and/or their bacteriocins, alone or in combination with other strategies) are also reported. Undoubtedly and considering the significant risk posed by this pathogen, especially against vulnerable population groups (e.g., younger, oldest, pregnant and immunocompromised), the better understanding of the various genetic and physiological underlying mechanisms leading to its antimicrobial recalcitrance, together with the influence of pre-existing resident/transient microbiota on its sessile behavior, is significant towards our efforts to develop fast, efficient, safe, and cost-effective prevention and control treatments to improve the safety of the food supply.

The role of biofilms in the development and dissemination of microbial resistance within the food industry is surely important and multifaceted. The articles presented in this Special Issue aim to contribute to understand this problem and its magnitude, making clear the need for novel efficient intervention methods.

Author Contributions: All authors have made a substantial, direct, and intellectual contribution to the work and approved it for publication. All authors have read and agreed to the published version of the manuscript.

Funding: This research received no external funding.

Conflicts of Interest: The authors declare no conflict of interest.

References

1. Bridier, A.; Briandet, R.; Thomas, V.; Dubois-Brissonnet, F. Resistance of bacterial biofilms to disinfectants: A review. *Biofouling* **2011**, *27*, 1017–1032. [CrossRef] [PubMed]
2. Giaouris, E.; Simões, M. Pathogenic biofilm formation in the food industry and alternative control strategies. In *Handbook of Food Bioengineering, Foodborne Diseases*; Holban, A.M., Grumezescu, A.M., Eds.; Academic Press (Elsevier): Amsterdam, The Netherlands, 2018; Volume 15, Chapter 11; pp. 309–377. [CrossRef]
3. Ostrov, I.; Paz, T.; Shemesh, M. Robust biofilm-forming *Bacillus* isolates from the dairy environment demonstrate an enhanced resistance to cleaning-in-place procedures. *Foods* **2019**, *8*, 134. [CrossRef] [PubMed]
4. Wallis, J.K.; Krömker, V.; Paduch, J.-H. Biofilm challenge: Lactic acid bacteria isolated from bovine udders versus staphylococci. *Foods* **2019**, *8*, 79. [CrossRef] [PubMed]

5. Sharma, K.; Pagedar Singh, A. Antibiofilm effect of DNase against single and mixed species biofilm. *Foods* **2018**, *7*, 42. [CrossRef] [PubMed]
6. Giaouris, E.; Nesse, L.L. Attachment of *Salmonella* spp. to food contact and product surfaces and biofilm formation on them as stress adaptation and survival strategies. In *Salmonella: Prevalence, Risk Factors and Treatment Options*; Hackett, C.B., Ed.; Nova Science Publishers, Inc.: New York, NY, USA, 2015; Chapter 6; pp. 111–136.
7. Paz-Méndez, A.M.; Lamas, A.; Vázquez, B.; Miranda, J.M.; Cepeda, A.; Franco, C.M. Effect of food residues in biofilm formation on stainless steel and polystyrene surfaces by Salmonella enterica strains isolated from poultry houses. *Foods* **2017**, *6*, 106. [CrossRef] [PubMed]
8. Cadena, M.; Kelman, T.; Marco, M.L.; Pitesky, M. Understanding antimicrobial resistance (AMR) profiles of Salmonella biofilm and planktonic bacteria challenged with disinfectants commonly used during poultry processing. *Foods* **2019**, *8*, 275. [CrossRef] [PubMed]
9. Rodríguez-López, P.; Rodríguez-Herrera, J.J.; Vázquez-Sánchez, D.; López Cabo, M. Current knowledge on Listeria monocytogenes biofilms in food-related environments: Incidence, resistance to biocides, ecology and biocontrol. *Foods* **2018**, *7*, 85. [CrossRef] [PubMed]

Article

Robust Biofilm-Forming *Bacillus* Isolates from the Dairy Environment Demonstrate an Enhanced Resistance to Cleaning-in-Place Procedures

Ievgeniia Ostrov [1,2], Tali Paz [1] and Moshe Shemesh [1,*]

[1] Department of Food Sciences, Institute for Postharvest Technology and Food Sciences, Agricultural Research Organization (ARO), The Volcani Center, 7528809 Rishon LeZion, Israel; ievgenia.ostrov@mail.huji.ac.il (I.O.); talipaz@agri.gov.il (T.P.)

[2] The Hebrew University—Hadassah, 9112001 Jerusalem, Israel

* Correspondence: moshesh@agri.gov.il; Tel.: +972-39683868

Received: 27 January 2019; Accepted: 16 April 2019; Published: 20 April 2019

Abstract: One of the main strategies for maintaining the optimal hygiene level in dairy processing facilities is regular cleaning and disinfection, which is incorporated in the cleaning-in-place (CIP) regimes. However, a frail point of the CIP procedures is their variable efficiency in eliminating biofilm bacteria. In the present study, we evaluated the susceptibility of strong biofilm-forming dairy *Bacillus* isolates to industrial cleaning procedures using two differently designed model systems. According to our results, the dairy-associated *Bacillus* isolates demonstrate a higher resistance to CIP procedures, compared to the non-dairy strain of *B. subtilis*. Notably, the tested dairy isolates are highly persistent to different parameters of the CIP operations, including the turbulent flow of liquid (up to 1 log), as well as the cleaning and disinfecting effects of commercial detergents (up to 2.3 log). Moreover, our observations indicate an enhanced resistance of poly-γ-glutamic acid (PGA)-overproducing *B. subtilis*, which produces high amounts of proteinaceous extracellular matrix, to the CIP procedures (about 0.7 log, compared to the wild-type non-dairy strain of *B. subtilis*). We therefore suggest that the enhanced resistance to the CIP procedures by the dairy *Bacillus* isolates can be attributed to robust biofilm formation. In addition, this study underlines the importance of evaluating the efficiency of commercial cleaning agents in relation to strong biofilm-forming bacteria, which are relevant to industrial conditions. Consequently, we believe that the findings of this study can facilitate the assessment and refining of the industrial CIP procedures.

Keywords: dairy industry; biofilm; *Bacillus* species; biofilm derived spores; cleaning-in-place; disinfecting effect

---

1. Introduction

Microbial contamination, caused by biofilm-forming bacteria, is one of the main threats to the quality, safety, stability and nutritional value of dairy products [1,2]. Moreover, biofilms are not only a potential source of contamination; they can also increase the corrosion rate of equipment used in the milk industry, impair heat transfer, and increase fluid frictional resistance [3]. Therefore, controlling biofilm formation is of major importance to the dairy industry [4–6].

Members of the *Bacillus* genus are among the most commonly found biofilm-formers in dairy farms and processing plants [7–9]. In addition to aggressive biofilm, these bacteria are able to form heat-resistant endospores [10,11]. To this end, the biofilm matrix can serve as an epicenter for the ripening of spores, which can be released from it and cause continuous contamination of the production environment [12,13]. Spores, as well as biofilm cells, are highly resistant to antimicrobial agents, which makes it rather difficult to eliminate them [11,14]. Moreover, biofilm matrix offers additional protection for embedded endospores, allowing their survival and colonization in the surrounding environment, when conditions are favorable [15]. In *B. subtilis*, the matrix has two main components, an exopolysaccharide (EPS) and amyloid-like fibers. Another extracellular polymer, γ-poly-DL-glutamic acid (PGA), is produced in copious amounts by some *B. subtilis* strains [16–18].

The main strategy to prevent biofilm formation, applied in the dairy industry, is to clean and disinfect regularly before bacteria attach firmly to surfaces [19,20]. Cleaning and disinfection in dairy processing plants have been incorporated into the cleaning-in-place (CIP) regimes, which include regular cleaning of processing equipment with alkaline and acidic liquids at high temperatures and flow velocities [4,21,22]. However, a weak point of CIP processes, evident in both industrial- and laboratory-scale systems, is their variable efficiency in eliminating established biofilms [4,21,23]. It is conceivable that biofilm formation can facilitate bacterial adaptation and survival in certain environmental niches. We therefore hypothesized that aggressive biofilm formation by dairy-associated bacteria might increase their resistance to industrial cleaning procedures.

In the present study, we evaluated the susceptibility of strong biofilm-forming dairy *Bacillus* isolates to cleaning-in-place procedures using two different model systems, which resemble industrial cleaning conditions. Our results show that the dairy-associated *Bacillus* isolates demonstrate enhanced resistance to different aspects of the CIP procedures, including mechanical, chemo-biological and disinfecting effects. Such reduced susceptibility can be attributed to robust biofilm formation by the tested dairy *Bacillus*.

2. Materials and Methods

2.1. Bacterial Strains and Growth Conditions

The following bacterial strains were used in this study: (i) dairy-associated isolates, such as *B. paralicheniformis* S127 [24,25], *B. licheniformis* MS310, *B. subtilis* MS302, *B. paralicheniformis* MS303 [24]; (ii) non-dairy isolate *B. subtilis* NCIB3610 (descendant of *B. subtilis* Marburg); (iii) poly-γ-glutamic acid (PGA)-overproducing mutant derivatives of *B. subtilis* 3610, *B. subtilis* YC295 (Δ*ywcC*) and *B. subtilis* YY54 (Δ*pgdS*) (a gift of Y. Chai [18]). *B. licheniformis* MS310, *B. subtilis* MS302 and *B. paralicheniformis* MS303 whole-genome shotgun projects are deposited at DDBJ/EMBL/GenBank, under accession numbers MIPQ00000000, MIZD00000000, MIZE00000000 respectively.

For routine growth, the strains were propagated in Lysogeny broth (LB; 10 g tryptone, 5 g yeast extract, 5 g NaCl per liter, pH 7) or on a solidified LB medium, supplemented with 1.5% agar at 37 °C.

2.2. Generation of Biofilm-Derived Spores

Biofilm colonies were generated at 30 °C in a biofilm-promoting medium (LBGM = LB + 1% *v*/*v* glycerol + 0.1 mM MnSO$_4$) [26]. Biofilm-derived spores were obtained from colonies, as described previously [21]. Briefly, the grown (three-day-old) colonies, harvested and suspended in phosphate buffered saline (PBS; 0.01 M phosphate buffer, 0.0027 M KCl, 0.137 M NaCl per 200 mL, Sigma Aldrich, St. Louis, MO, USA), were disrupted by mild sonication (Vibra Cell, Sonics, Newtown, CT, USA; amplitude 60%, pulse 10 s, pause 10 s, duration 2 min, instrument power: 7.2 Joules per second). During sonication, the samples were kept on ice. Then, heat killing was performed at 80 °C for 20 min. Cell numbers after heat killing were quantified by the spread plating method.

2.3. Staining Extracellular Matrix of Biofilm-Derived Spores

Biofilm-derived spores were stained using the FilmTracer™ SYPRO® Ruby Biofilm Matrix Stain (Molecular Probes, Eugene, OR, USA), according to the manufacturer's protocol. Stained samples were visualized by confocal laser scanning microscopy (CLSM; Olympus IX81, Tokyo, Japan) at a 10 μm scale.

2.4. Preparation for Cleaning Tests and Enumeration of Biofilm-Derived Spores

The preparation of biofilm-derived spores for cleaning tests was performed, as described in the previous study [21]. Briefly, 200-μL aliquots of the spore suspension (containing approximately two million spores) were applied in the sampling area of stainless-steel sampling plates and dried in a biological laminar hood for 1 h. Two sampling plates were not exposed to the cleaning procedures (control). Following each cleaning test, the sampling plates were immediately subjected to abundant rinsing with tap water at RT (similar to the CIP procedures at Israeli dairy farms, where the rinsing with water stage is introduced after applying a cleaning agent). For the enumeration of the spores, the sampling area on each plate was carefully swabbed with cotton swabs, moistened in PBS buffer. Swabs from each plate were then agitated in PBS in separate test tubes. Serial dilutions from each sample were prepared, followed by spread plating on LB agar for CFU analysis. Plates were incubated for 24 h at 37 °C, before the colonies were counted. The efficiency of a cleaning procedure was evaluated by comparing the number of viable spores (attached to sampling plates), before and after cleaning.

2.5. Cleaning Solutions

The following cleaning solutions were used in this study: Caustic soda (NaOH), sodium hypochlorite (NaOCl) and six different commercial alkaline detergents, defined as solutions I (10–15% NaOH, 3–5% NaOCl), A (polycarboxylate, phosphates, 3.6% NaOCl), M (>5% polycarboxylate, 5–15% phosphates, 3.6% NaOCl), F (5% phosphonates, polycarboxylates), D (active chlorine, alkaline-based) and H (active chlorine, phosphates, additives, alkaline-based), which are commonly used in the Israeli dairy farms. The pH value of the tested solutions varied between 11–12; the pH of NaOH was 13; and the pH of NaOCl was 4. In accordance with the manufacturer's recommendations, the agents were used at the following concentrations: (i) 0.5% (*v/v*) for solutions A, M, F, D, H; (ii) 0.6% (*v/v*) for solution I; (iii) 0.5% (*m/v*) for caustic soda and detergent H; (iv) 0.018% (*v/v*) for sodium hypochlorite (similar to the NaOCl concentration in working solutions of the examined cleaning agents, such as A, M and I). As a control, tap water was used (pH value around 7.7), with a standard level of hardness (50 mg/L Ca^{2+}, 50 mg/L Mg^{2+}), without the addition of any detergent.

2.6. Cleaning Test Installations

The cleaning tests were carried out either using the cleaning-in-place (CIP) model system (closely resembling the typical conditions for milking systems) [21] or using the simplified laboratory procedure, developed in this study.

2.6.1. CIP Model System

The main components of the CIP model system were described in the previous study [21]. In brief, the system consists of a 5-m stainless-steel milk line (fitted with a test unit) for pumping the cleaning agents from the basin, milk releaser, and a stainless-steel return line to the basin. The test unit has T-junctions, protruding 35, 125 or 275 mm from the main loop, reflecting different degrees of cleaning difficulty. Sampling plates with the spores were mounted on the T-junctions and cleaned in the installation. The temperature of the cleaning solution during the cleaning tests was 50 °C. To generate flushing pulsation of the circulating liquid, air was introduced into the system every 8 s. The duration of each cleaning cycle was 10 min.

2.6.2. Laboratory System

For cleaning tests in the laboratory system, sampling plates with the spores were placed into 100 mL plastic vessels (Yoel Naim, Rehovot, Israel), containing 50 mL of cleaning solution (preliminarily warmed to 50 °C). The samples were incubated in closed vessels at conditions simulating those in the CIP-model system (50 °C, 250 rpm) for 10 min.

2.7. Evaluation of the Effect of the Cleaning Agents on the Viability of Bacillus Spores

The tested solutions were added to spore suspension within tap water containing around 1×10^7 CFU/mL spores. The spore suspension without the addition of detergents was used as a control. The samples were incubated in closed tubes under the conditions of the laboratory system (50 °C, 250 rpm) for 10 min. The CFU measurements of the number of viable spores were made immediately after the addition of the tested cleaning agents and following 10 min of incubation.

2.8. Statistical Analysis

The results of the study are the means and standard deviation (SD) of at least two independent biological experiments, performed in triplicate. The Student's *t* test was used to calculate the significance of the difference between the mean expression of a given experimental sample and the control sample. A *p* value of <0.05 was considered significant.

3. Results

3.1. Dairy-Associated Bacillus Isolates Exhibit Robust Biofilm Phenotype Compared to B. subtilis 3610

We focused this investigation on biofilm-forming milk isolates of *Bacillus* species, which were obtained from Israeli dairy farms and recently identified and characterized [24]. The isolates were further characterized using a colony-type biofilm model for the robustness of their biofilm-forming capabilities (Figure 1; Table S1). We found notable differences in the colony-biofilm phenotype between *B. subtilis* 3610 and the dairy *Bacillus* isolates (Figure 1A). Thus, the biofilm colonies of *B. subtilis* 3610 had a complex "wrinkled" structure (shown to be a network of channels rich in biofilm matrix-producing cells [27,28]), but were not mucoid. The colonies of the tested dairy-associated strains combined an intricate "wrinkled" phenotype with the formation of highly mucoid "channel"- and "ridge"-like structures, not observed for *B. subtilis* 3610 (Figure 1A).

To support this observation, we analyzed the extracellular matrix content in the colony biofilm of the tested dairy *Bacillus* isolates and *B. subtilis* 3610 by visualizing matrix proteins. Our results indicate that biofilm cells/spores, harvested from colonies of the dairy-associated strains, could be surrounded by higher amounts of extracellular polymeric substances (EPS), compared to *B. subtilis* 3610 (Figure 1B).

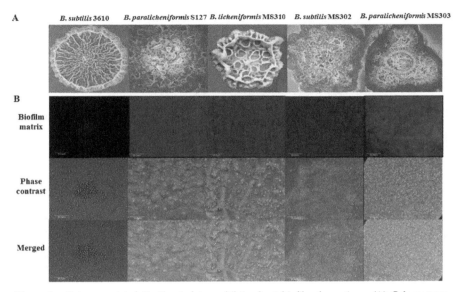

Figure 1. Dairy-associated *Bacillus* isolates exhibit robust biofilm formation. (**A**) Colony type biofilm formation by the tested *Bacillus* strains in the biofilm-promoting medium, LBGM. The images were taken using a stereoscopic microscope (Zeiss Stemi 2000-C; Carl Zeiss, Gottingen, Germany). (**B**) Biofilm-derived spores of the dairy *Bacillus* strains are surrounded by high amounts of the extracellular matrix. Protein components of the biofilm matrix were stained red. The samples were analyzed using a confocal laser scanning microscope (CSLM, Olympus, Japan). Scale: 10 μm.

3.2. Dairy-Associated Bacillus Isolates Display an Enhanced Resistance to the Mechanical Effect of Water Circulation

Primarily, we evaluated the susceptibility of the tested strains to water circulation in the CIP model system (closely resembling the conditions typical for milking pipes). Cleaning with water alone reflects the mechanical cleaning effect brought about by the flow of liquid in the installation [21,29]. The susceptibility of the dairy-associated *Bacillus* strains to cleaning procedures was compared to the non-dairy isolate *B. subtilis* 3610 (used as a model strain in our previous study [20]). In order to simulate dairy biofilm, we used a system that is based on the biofilm-derived spores of the tested *Bacillus*, obtained from the biofilm colonies as previously described [21].

We found that the biofilm-derived spores of the dairy *Bacillus* were significantly (by 0.3–1 log) more resistant to water circulation, compared to *B. subtilis* 3610, in the case of 35 and 125 mm T-junctions (representing high levels of turbulence; Figure 2). In the samples placed into the 275-mm T-junctions (the lowest degree of turbulence available in the CIP model system), the susceptibility to cleaning was either similar (*B. paralicheniformis* S127) or lower by 0.1–0.3 log (*B. paralicheniformis* MS303, *B. licheniformis* MS310, *B. subtilis* MS302) than the control samples.

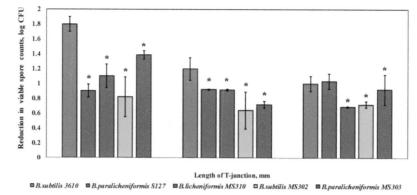

Figure 2. Effect of the cleaning procedure with tap water on the removal of biofilm-derived spores of the dairy-associated *Bacillus* in the CIP model system. Sampling plates, each containing approximately 2 million spores of *B. subtilis* 3610 or dairy *Bacillus* isolates, were mounted on T-junctions, protruding 35, 125, and 275 mm from the main loop of the CIP model system, and cleaned in the installation. Tap water, without the addition of any detergent, was used as the cleaning agent. A basic assumption was the similar adhesion efficiency of the spores of each tested strain in different experimental repeats (since the spores were obtained using previously validated experimental procedures [21]). The cleaning effect was evaluated by comparing the number of viable spores (attached to the sampling plates), before and after cleaning. The results represent the means and standard deviations (SD) of two independent biological experiments, performed in triplicate. * Statistically significant difference ($p < 0.05$) between the reduction in the viable spore counts of a given sample and the reduction in the spore counts for *B. subtilis* 3610 (control).

Next, we wanted to test the persistence of the examined *Bacillus* strains against the chemical effect of the commercial cleaning solutions. Since the chemical effect of the cleaning agents is less dependent on the flow turbulence, it was decided to simplify our experimental system to a lab-scale cleaning test (hereinafter referred to as the laboratory system). We first confirmed the validity of this system by comparing the strains' ability to withstand a mechanical effect. Importantly, the dairy-associated *Bacillus* demonstrated an enhanced resistance to water circulation (by 0.6–0.7 log), compared to *B. subtilis* 3610, also during the cleaning tests performed in the laboratory system (Figure S1). A strong correlation between the results obtained in the two differently designed experimental systems indicates the reliability of the approach used.

3.3. Dairy-Associated Bacillus Isolates Demonstrate an Enhanced Resistance to Commercial Cleaning Agents during CIP Procedures

Next, we evaluated the susceptibility to commercial cleaning agents of two selected dairy-associated isolates, *B. paralicheniformis* S127 and *B. licheniformis* MS310, which demonstrated the highest amount of EPS surrounding biofilm bacteria, according to a relative fluorescence analysis, in comparison to *B. subtilis* 3610 (Table S1). Consequently, we performed cleaning procedures using six different alkaline detergents, caustic soda (NaOH) and sodium hypochlorite (NaOCl) at concentrations recommended by the manufacturers. It was found that *B. licheniformis* MS310, as well as *B. paralicheniformis* S127, were more resistant to the tested solutions (up to 2.3 and 0.76 log, respectively), compared to *B. subtilis* 3610 (Figure 3). Interestingly, *B. subtilis* 3610 was particularly susceptible to agents I, M, D and H, whereas *B. paralicheniformis* S127 was highly persistent to cleaning with agent H and NaOH, but similarly susceptible to solutions I, M and F as *B. subtilis* 3610. *B. licheniformis* MS310 was exceedingly resistant to treatment by the examined solutions, especially to agents I, M and H (Figure 3).

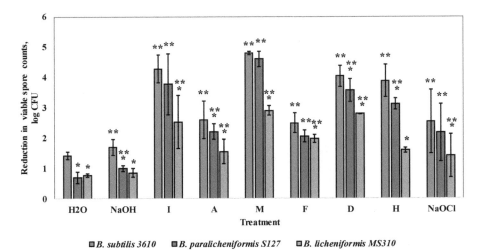

Figure 3. Effect of commercial cleaning agents on the removal of the biofilm-derived spores of the dairy-associated *Bacillus* in the simplified laboratory system. Sampling plates, each maintaining approximately 2 million spores of the tested *Bacillus* strains, were cleaned in the laboratory system. Caustic soda, sodium hypochlorite and the following cleaning solutions—I, A, M, F, D and H (compositions and dosages are described in Methods)—were used as the cleaning agents. The cleaning effect was evaluated by comparing the numbers of viable spores (attached to sampling plates), before and after cleaning. The results represent the means and standard deviation (SD) of two independent biological experiments, performed in triplicate. * Statistically significant difference ($p < 0.05$) between the reduction in the viable spore counts in a given sample and the reduction in the spore counts for *B. subtilis* 3610 (control). ** Statistically significant difference ($p < 0.05$) between the reduction in the viable spore counts, after treatment with a given cleaning agent, and the reduction in the spore counts for the same strain, after incubation with tap water.

As indicated in the previous study [21], the biofilm removal effect of a cleaning agent includes both the mechanical effect of the liquid circulation and the chemo-biological effect from the active components, present in the agent. To gain greater insight into the mode of action of the examined solutions, we calculated their chemo-biological effect in relation to the biofilm-derived spores of the tested strains. As shown in Figure S2, *B. licheniformis* MS310 was significantly more resistant to the chemo-biological effect of the examined solutions, compared to the other strains. At the same time, in most cases, *B. paralicheniformis* S127 was equally susceptible to the chemo-biological effect, compared to 3610. This indicates that the tested strains have varying degrees of resistance to the mechanical and chemo-biological effects of cleaning agents. Thus, the low susceptibility of MS310 to the examined solutions results from the increased resistance both to their mechanical and chemo-biological effect (Figure S3). In the case of S127, a high resistance to the majority of the tested solutions (NaOH, I, F, D) is caused mainly by the low susceptibility to the mechanical removal of spores, while the persistence to agents A and H results from a reduced sensitivity to both the mechanical and chemo-biological impacts (Figure 3; Figure S3).

3.4. Dairy-Associated Bacillus Isolates Demonstrate an Enhanced Resistance to the Disinfecting Effect of the Tested Agents

Primarily, we determined the ability of the tested agents to remove surface-attached spores, without affecting the viability (cleaning effect) and/or inactivating the spores (disinfecting effect). For this, spore suspensions were incubated with each of the tested agents under the conditions of the laboratory system. We found that the examined agents had different influences on the viability of the biofilm-derived spores of the tested strains (Figure 4). Thus, solutions D and M notably reduced the spore counts of *B. subtilis* 3610, after 10 min of incubation (Figure 4); there was a 0.5 log reduction in the viable spores for S127, after incubation with solution I; while none of the tested solutions affected the viability of the MS310 spores. Interestingly, NaOCl, commonly used as a disinfecting agent, did not influence the viability of the tested strains at the examined concentration (the dosage widely used in industrial cleaning agents; Figure 4).

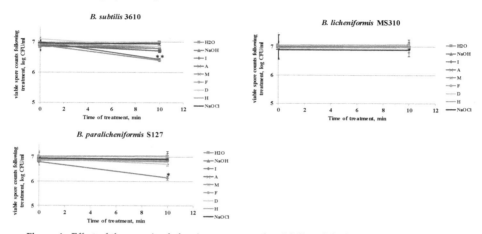

Figure 4. Effect of the examined cleaning agents on the viability of the biofilm-derived spores of the tested *Bacillus* strains. Caustic soda, sodium hypochlorite, and different cleaning solutions—I, A, M, F, D, and H (compositions are described in Methods)—were added to the tubes, with spore suspension of the tested *Bacillus* isolates. Spore suspension, without any detergent, was used as the control. The effect on spore viability was evaluated by comparing the numbers of viable spores in the control and after the treatment with the tested agents (following 10 min of incubation at 50 °C, 250 rpm). The results represent the means and standard deviation (SD) of two independent biological experiments, performed in duplicate. * Statistically significant difference ($p < 0.05$) between the viable spore counts in a given sample versus the spore counts after cleaning with water (control).

Next, we determined a correlation between the cleaning and disinfecting effects of the tested detergents. Thus, we defined the ability of a cleaning agent to reduce the number of viable spores after 10 min of a cleaning cycle, as a disinfecting effect. We compared the percentage of the disinfecting effect to the total chemo-biological effect of a cleaning agent (taken as 100%). The difference between the total chemo-biological effect of the tested agent and the disinfecting effect was defined as the cleaning effect [21]. As can be inferred from Figure 5, the ratio between the cleaning and disinfecting effects of the examined detergents differed for the tested strains. Thus, the removal of the MS310 spores was due solely to the cleaning effect of the tested solutions. *B. paralicheniformis* S127 was significantly more resistant to the disinfecting effect of agents A, M, F, H, and NaOH, compared to *B. subtilis* 3610, but much more susceptible to the disinfecting effect of solution I (Figure 5). Overall, the chemo-biological effect of the tested agents was mostly due to the removal of surface-attached spores (cleaning effect) and not to disinfecting.

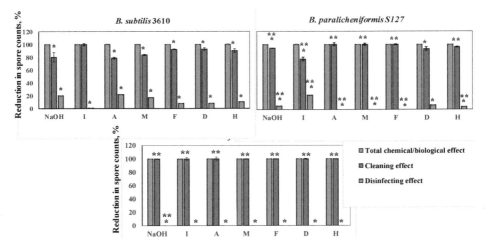

Figure 5. Correlation between the cleaning and disinfecting effects of the examined agents for each tested strain. Caustic soda and different cleaning solutions—I, A, M, F, D, and H (compositions are described in Methods)—were added to the tubes, with spore suspension, of the tested *Bacillus* isolates and incubated for 10 min at 50 °C, 250 rpm. The ability of a cleaning agent to reduce the number of viable spores was defined as the disinfecting effect. The percentage of the disinfecting effect was compared to the total chemical/biological effect of a cleaning agent (taken as 100%). The difference between the total chemical/biological effect of a cleaning agent and the disinfecting effect was defined as the cleaning effect. The results represent the means and standard deviation (SD) of two independent biological experiments, performed in duplicate. * Statistically significant difference ($p < 0.05$) between the reduction in the spore counts due to the cleaning or disinfecting effects versus the total chemo-biological effect of a tested agent. ** Statistically significant difference ($p < 0.05$) between the reduction in the viable spore counts in a given sample and the reduction in the spore counts for *B. subtilis* 3610 (control).

4. Discussion

It becomes increasingly clear that biofilm formation by *Bacillus* species can facilitate their survival in the dairy environment [11,21]. Our current study investigated the effect of CIP procedures on strong biofilm-forming dairy *Bacillus*, compared to the non-dairy *B. subtilis* 3610, using differently designed model systems. As in our previous study [21], we used biofilm-derived spores to simulate the type of hygiene problem common in practice. Thus, similarly to actual dairy biofilm, biofilm-derived spores combine the presence of biofilm matrix [21] and a high content of spores [29,30]. Moreover, the resistance of vegetative cells/spores to cleaning and disinfection can be greatly enhanced by the presence of EPS [21,31]. At the same time, the presence of spores within the *Bacillus* biofilm may also modify biofilm properties, e.g., interaction forces [12].

In the current study, two model systems were used to ensure that the enhanced resistance of the dairy isolates to cleaning procedures is observed under different experimental conditions, which are relevant to the industrial CIP systems. Moreover, the design of the CIP system, employed in our previous study does not allow for the evaluation of the disinfecting effect of the cleaning agents on *Bacillus* spores directly in this system [21]. The laboratory system, developed in this study, provides sufficient conditions both for determining the mechanical, chemo-biological and disinfecting effects of the cleaning agents.

A first notable finding of the study was the enhanced resistance of the dairy *Bacillus* to the mechanical effect of liquid circulation. Thus, the most expressed difference in cleaning susceptibility between the dairy-associated strains and *B. subtilis* 3610 was observed at high levels of turbulence (35- and 125-mm T-junctions, CIP model system; Figure 2). In the case of a lower turbulence (275-mm

T-junction), the difference between the dairy *Bacillus* isolates and the non-dairy strain is markedly decreased, and for some strains, it was insignificant (Figure 2). These results suggest that the protective effect of *Bacillus* biofilm matrix is most strongly expressed under a high turbulence of liquid flow. Previous studies demonstrate that a high turbulence may facilitate the removal of surface-attached bacteria [21,32–34], but may also increase the rate of attachment by bringing the microbial cells and the substrate in close proximity [35]. Thus, biofilm formation by the dairy-associated *Bacillus* can be detrimental not only in so-called "dead legs" (equipment details, in which the flow of liquid is significantly less turbulent), but also in main pipelines.

Furthermore, we showed that the biofilm-derived spores of the dairy *Bacillus* isolates are much more resistant to commercial cleaning agents, compared to *B. subtilis* 3610. Presumably, the causes of this resistance differ between the tested strains. Thus, the biofilm-derived spores of MS310 are, apparently, less susceptible both to the mechanical and chemo-biological effects of the employed solutions (Figures S2 and S3). At the same time, *B. paralicheniformis* S127 has the highest resistance to the mechanical removal of spores but shows a variable susceptibility to the chemo-biological effect of the tested agents.

As shown in our previous study [21], the chemo-biological effect of cleaning agents comprises a disinfecting effect (inactivating bacteria) and/or removal of them from the surfaces of dairy equipment (cleaning effect). According to our results, the dairy *Bacillus* isolates are significantly less susceptible to the disinfecting effect of the tested agents, compared to the non-dairy strain (except solution I in the case of S127; Figure 4; Figure 5). The observed differences in the mechanical and chemo-biological effects between the tested strains might be explained by the dissimilarities in the biofilm structure. For instance, a correlation between colony biofilm phenotype of the tested strains, and their resistance to the cleaning procedures, was observed (Figure 1). Thus, the dairy-associated *Bacillus*, characterized by a mucoid biofilm phenotype, were less susceptible to mechanical and chemo-biological effects during the CIP procedures. Since biofilm matrix components can be responsible for binding and/or neutralizing detergents and antimicrobial agents [36,37], differences in the matrix structure/composition can lead to differences in cleaning and/or disinfection susceptibility. Thereby, the biofilm matrix composition was shown to affect the susceptibility of food-associated staphylococci to cleaning and disinfection agents, with polysaccharide matrix-producing strains being more resistant to the lethal effect of benzalkonium chloride [38]. Likewise, the efficiency of monochloramine disinfection was dependent on the quantity and composition of EPS in *Pseudomonas* biofilms. Protein-based EPS-producing *P. putida* was less sensitive to monochloramine than polysaccharide-based EPS-producing *P. aeruginosa*, since monochloramine had a selective reactivity with proteins over polysaccharides [39]. According to Bridier et al. (2011) [40], the biofilm of the *P. aeruginosa* clinical isolate, in which a high delay of benzalkonium chloride penetration is recorded, was characterized by a large quantity of proteinacious matrix. Moreover, the authors report that, in *P. aeruginosa*, resistance to antimicrobial agents is intimately related to the inherent three-dimensional organization of cells into the exopolymeric matrix. Therefore, the low sensitivity of the dairy *Bacillus* isolates to the CIP procedures (compared to *B. subtilis* 3610) may be connected to differences in the structure/composition of the biofilm matrix.

Importantly, mucoid colony formation, observed for the dairy *Bacillus* isolates, was viewed as a hallmark of poly-γ-glutamic acid (PGA) production in multiple previous studies [17,18]. Significant production of PGA could result in a stronger attachment to surfaces due to its adhesive properties [41]. To this end, PGA-overproducing derivatives of *B. subtilis* 3610 (*B. subtilis* YC295 and *B. subtilis* YY54) were significantly more resistant to the mechanical effect of water circulation, compared to the wild type (Figure 6C). Notably, biofilm colonies of these mutant strains were more mucoid, compared to the WT (Figure 6A). Moreover, the biofilm-derived spores of PGA-overproducing *B. subtilis* were surrounded by higher amounts of proteinaceous extracellular matrix, which resembles the tested dairy *Bacillus* isolates (Figure 6B). Therefore, the presence of PGA in the biofilm matrix of the examined bacterial strains may be one of the factors enhancing resistance to the CIP procedures.

We believe that the role of PGA and other presumptive EPS components of the dairy-associated *Bacillus* in relation to cleaning and disinfecting agents is an important subject for further investigation.

Relatively low cleaning and, especially, disinfecting effects of the tested solutions (Figure 5) might lead to undesirable implications regarding the hygiene level in dairy environments. For instance, the rapid recovery of biofilms after inappropriate disinfectant treatment is often observed. This may be due to the re-growth of surviving cells, residual biofilm, providing a conditioning layer for further cell attachment, or the selection of resistant microorganisms that survive and thrive after antimicrobial treatment [5]. In addition, biofilm cells exposure to low (sub-lethal) concentrations of disinfecting compounds, including chlorine-based detergents, can stimulate further biofilm development [10,42,43]. Therefore, we speculate that the composition of commercial CIP agents should be revised and evaluated under the experimental conditions suggested in this study.

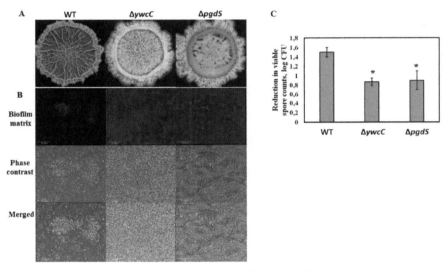

Figure 6. PGA-overproducing derivatives of *B. subtilis* 3610 exhibit increased resistance to the CIP procedures due to enhanced biofilm formation. (**A**) Colony biofilm formation by the tested *Bacillus* strains in the biofilm-promoting medium, LBGM. The images were taken using a stereoscopic microscope (Zeiss Stemi 2000-C; Carl Zeiss, Gottingen, Germany). (**B**) Biofilm-derived spores of the PGA-overproducing *B. subtilis* strains are surrounded by high amounts of extracellular matrix. Protein components of the biofilm matrix were stained red. The samples were analyzed using a confocal laser scanning microscope (CSLM, Olympus, Japan). Scale: 10 μm. (**C**) The effect of water circulation on the removal of biofilm-derived spores of the PGA-overproducing derivatives of *B. subtilis* 3610 in the laboratory CIP system. * Statistically significant difference ($p < 0.05$) between the reduction in the viable spore counts in a given sample and the reduction in the spore counts for *B. subtilis* 3610 (control).

5. Conclusions

We demonstrated in this study that the dairy-associated *Bacillus* isolates are characterized by an enhanced resistance to different aspects of the CIP procedures, such as the mechanical, chemo-biological, and disinfecting effects, compared to the non-dairy *Bacillus*. Such increased resistance can be attributed to robust biofilm formation by the tested dairy *Bacillus*. The results of the study underline the importance of revising the composition of commercial cleaning agents and evaluating their efficiency in relation to strong biofilm-forming bacteria, relevant to industrial conditions. To this end, the biofilm-derived spores of the dairy-associated *Bacillus*, examined in this study, can be used as an appropriate model for assessing and refining the CIP procedures.

Supplementary Materials: The following are available online at http://www.mdpi.com/2304-8158/8/4/134/s1, Figure S1: Effect of the cleaning procedure with tap water on removal of the biofilm-derived spores of the dairy-associated *Bacillus* in the simplified laboratory system, Figure S2: Chemo-biological effect of the commercial cleaning agents on removal of the biofilm derived spores in the laboratory CIP system, Figure S3: Correlation between mechanical and chemo-biological effect of the examined agents in relation to the removal of the biofilm derived spores in the laboratory CIP system, Table S1: Relative quantity of the matrix, surrounding biofilm-derived spores of the dairy-associated *Bacillus* isolates and *B. subtilis* 3610.

Author Contributions: Conceptualization, I.O. and M.S.; methodology, I.O.; investigation, I.O. and T.P.; data curation, I.O.; writing—original draft preparation, I.O.; writing—review & editing, M.S.; supervision, M.S.; funding acquisition, M.S.

Acknowledgments: We thank Doron Steinberg from The Hebrew University of Jerusalem for the helpful discussions. We thank Golan Yakov, Avraham Harel and Eduard Belausov for excellent technical assistance. We are also grateful to Yulia Kroupitski, Rama Falk and Adin Shwimmer for their supportive suggestions and discussions. This work was partially supported by the Israel Dairy Board [grant number 4210343].

Conflicts of Interest: The authors declare no conflict of interest.

References

1. Ivy, R.A.; Ranieri, M.L.; Martin, N.H.; den Bakker, H.C.; Xavier, B.M.; Wiedmann, M.; Boor, K.J. Identification and Characterization of Psychrotolerant Sporeformers Associated with Fluid Milk Production and Processing. *Appl. Environ. Microbiol.* **2012**, *78*, 1853–1864. [CrossRef]

2. Ranieri, M.L.; Huck, J.R.; Sonnen, M.; Barbano, D.M.; Boor, K.J. High temperature, short time pasteurization temperatures inversely affect bacterial numbers during refrigerated storage of pasteurized fluid milk. *J. Dairy Sci.* **2009**, *92*, 4823–4832. [CrossRef] [PubMed]

3. Kumar, C.G.; Anand, S.K. Significance of Microbial Biofilms in Food Industry: A Review. *Int. J. Food Microbiol.* **1998**, *42*, 9–27. [CrossRef]

4. Bremer, P.J.; Fillery, S.; McQuillan, A.J. Laboratory Scale Clean-In-Place (CIP) Studies on the Effectiveness of Different Caustic and Acid Wash Steps on the Removal of Dairy Biofilms. *Int. J. Food Microbiol.* **2006**, *106*, 254–262. [CrossRef] [PubMed]

5. Flint, S.H.; Bremer, P.J.; Brooks, J.D. Biofilms in Dairy Manufacturing Plant-Description, Current Concerns and Methods of Control. *Biofouling* **1997**, *11*, 81–97. [CrossRef]

6. Pasvolsky, R.; Zakin, V.; Ostrova, I.; Shemesh, M. Butyric Acid Released during Milk Lipolysis Triggers Biofilm Formation of *Bacillus* species. *Int. J. Food Microbiol.* **2014**, *181*, 19–27. [CrossRef] [PubMed]

7. Sharma, M.; Anand, S.K. Bacterial Biofilm on Food Contact Surfaces: A Review. *J. Food Sci. Technol.* **2002**, *39*, 573–593.

8. Sharma, M.; Anand, S.K. Biofilms Evaluation as an Essential Component of HACCP for Food/Dairy Processing Industry—A Case. *Food Control* **2002**, *13*, 469–477. [CrossRef]

9. Sharma, M.; Anand, S.K. Characterization of Constitutive Microflora of Biofilms in Dairy Processing Lines. *Food Microbiol.* **2002**, *19*, 627–636. [CrossRef]

10. Burgess, S.A.; Lindsay, D.; Flint, S.H. Thermophilic Bacilli and their Importance in Dairy Processing. *Int. J. Food Microbiol.* **2010**, *144*, 215–225. [CrossRef] [PubMed]

11. Shaheen, R.; Svensson, B.; Andersson, M.A.; Christiansson, A.; Salkinoja-Salonen, M. Persistence strategies of *Bacillus cereus* spores isolated from dairy silo tanks. *Food Microbiol.* **2010**, *27*, 347–355. [CrossRef] [PubMed]

12. Ryu, J.H.; Beuchat, L.R. Biofilm Formation and Sporulation by *Bacillus cereus* on a Stainless Steel Surface and Subsequent Resistance of Vegetative Cells and Spores to Chlorine, Chlorine Dioxide, and a Peroxyacetic Acid-Based Sanitizer. *J. Food Prot.* **2005**, *68*, 2614–2622. [CrossRef]

13. Wijman, J.G.; de Leeuw, P.P.; Moezelaar, R.; Zwietering, M.H.; Abee, T. Air-liquid Interface Biofilms of *Bacillus cereus*: Formation, Sporulation, and Dispersion. *Appl. Environ. Microbiol.* **2007**, *73*, 1481–1488. [CrossRef] [PubMed]

14. Faille, C.; Benezech, T.; Blel, W.; Ronse, A.; Ronse, G.; Clarisse, M.; Slomianny, C. Role of Mechanical vs. Chemical Action in the Removal of Adherent *Bacillus* Spores during CIP Procedures. *Food Microbiol.* **2013**, *33*, 149–157. [CrossRef] [PubMed]

15. Branda, S.S.; Gonzalez-Pastor, J.E.; Ben-Yehuda, S.; Losick, R.; Kolter, R. Fruiting Body Formation by *Bacillus subtilis*. *Proc. Natl. Acad. Sci. USA* **2001**, *98*, 11621–11626. [CrossRef] [PubMed]

16. Morikawa, M.; Kagihiro, S.; Haruki, M.; Takano, K.; Branda, S.; Kolter, R.; Kanaya, S. Biofilm Formation by a *Bacillus subtilis* Strain that Produces Gamma-Polyglutamate. *Microbiol. Sgm* **2006**, *152*, 2801–2807. [CrossRef]
17. Stanley, N.R.; Lazazzera, B.A. Defining the Genetic differences Between Wild and Domestic Strains of *Bacillus subtilis* that Affect Poly-gamma-DL-glutamic Acid Production and Biofilm Formation. *Mol. Microbiol.* **2005**, *57*, 1143–1158. [CrossRef]
18. Yu, Y.Y.; Yan, F.; Chen, Y.; Jin, C.; Guo, J.H.; Chai, Y.R. Poly-gamma-Glutamic Acids Contribute to Biofilm Formation and Plant Root Colonization in Selected Environmental Isolates of *Bacillus subtilis*. *Front. Microbiol.* **2016**, *7*, 1811. [CrossRef]
19. Midelet, G.; Carpentier, B. Impact of Cleaning and Disinfection Agents on Biofilm Structure and on Microbial Transfer to a Solid Model Food. *J. Appl. Microbiol.* **2004**, *97*, 262–270. [CrossRef]
20. Simoes, M.; Simoes, L.C.; Vieira, M.J. A Review of Current and Emergent Biofilm Control Strategies. *Lwt-Food Sci. Technol.* **2010**, *43*, 573–583. [CrossRef]
21. Ostrov, I.; Harel, A.; Bernstein, S.; Steinberg, D.; Shemesh, M. Development of a Method to Determine the Effectiveness of Cleaning Agents in Removal of Biofilm Derived Spores in Milking System. *Front. Microbiol.* **2016**, *7*, 1498. [CrossRef] [PubMed]
22. Zottola, E.A.; Sasahara, K.C. Microbial Biofilms in the Food-Processing Industry-Should They Be a Concern. *Int. J. Food Microbiol.* **1994**, *23*, 125–148. [CrossRef]
23. Faille, C.; Fontaine, F.; Benezech, T. Potential Occurrence of Adhering Living *Bacillus* Spores in Milk Product Processing Lines. *J. Appl. Microbiol.* **2001**, *90*, 892–900. [CrossRef] [PubMed]
24. Ostrov, I.; Sela, N.; Belausov, E.; Steinberg, D.; Shemesh, M. Adaptation of *Bacillus* Species to Dairy Associated Environment Facilitates their Biofilm Forming Ability. *Food Microbiol.* **2019**. [CrossRef]
25. Ostrov, I.; Sela, N.; Freed, M.; Khateb, N.; Kott-Gutkowski, M.; Inbar, D.; Shemesh, M. Draft Genome Sequence of *Bacillus licheniformis* S127, Isolated from a Sheep Udder Clinical Infection. *Genome Announc.* **2015**, *3*. [CrossRef] [PubMed]
26. Shemesh, M.; Chai, Y.R. A Combination of Glycerol and Manganese Promotes Biofilm Formation in *Bacillus subtilis* via Histidine Kinase KinD Signaling. *J. Bacteriol.* **2013**, *195*, 2747–2754. [CrossRef] [PubMed]
27. Bridier, A.; Tischenko, E.; Dubois-Brissonnet, F.; Herry, J.M.; Thomas, V.; Daddi-Oubekka, S.; Waharte, F.; Steenkeste, K.; Fontaine-Aupart, M.P.; Briandet, R. Deciphering Biofilm Structure and Reactivity by Multiscale Time-Resolved Fluorescence Analysis. *Bact. Adhes. Chem. Biol. Phys.* **2011**, *715*, 333–349. [CrossRef]
28. Vlamakis, H.; Chai, Y.R.; Beauregard, P.; Losick, R.; Kolter, R. Sticking Together: Building a Biofilm the *Bacillus subtilis* Way. *Nat. Rev. Microbiol.* **2013**, *11*, 157–168. [CrossRef] [PubMed]
29. Sundberg, M.; Christiansson, A.; Lindahl, C.; Wahlund, L.; Birgersson, C. Cleaning Effectiveness of Chlorine-free Detergents for Use on Dairy Farms. *J. Dairy Res.* **2011**, *78*, 105–110. [CrossRef]
30. Faille, C.; Benezech, T.; Midelet-Bourdin, G.; Lequette, Y.; Clarisse, M.; Ronse, G.; Ronse, A.; Slomianny, C. Sporulation of *Bacillus* spp. within Biofilms: A Potential Source of Contamination in Food Processing Environments. *Food Microbiol.* **2014**, *40*, 64–74. [CrossRef] [PubMed]
31. Xue, Z.; Sendamangalam, V.R.; Gruden, C.L.; Seo, Y. Multiple Roles of Extracellular Polymeric Substances on Resistance of Biofilm and Detached Clusters. *Environ. Sci. Technol.* **2012**, *46*, 13212–13219. [CrossRef]
32. Leliévre, C.; Antonini, G.; Faille, C.; Bénézech, T. Cleaning-in-place, Modelling of Cleaning Kinetics of Pipes Soiled by *Bacillus* Spores Assuming a Process Combining Removal and Deposition. *Food Bioprod. Process.* **2002**, *80*, 305–311. [CrossRef]
33. Leliévre, C.; Legentilhomme, P.; Legrand, J.; Faille, C.; Bénézech, T. Hygenic Design: Influence of the Local Wall Shear Stress Variations on the Cleanability of a Three-way Valve. *Chem. Eng. Res. Des.* **2003**, *81*, 1071–1076. [CrossRef]
34. Wirtanen, G.; Husmark, U.; Mattila-Sandholm, T. Microbial Evaluation of the Biotransfer Potential from Surfaces with *Bacillus* Biofilms after Rinsing and Cleaning Procedures in Closed Food-Processing Systems. *J. Food Prot.* **1996**, *59*, 727–733. [CrossRef]
35. Palmer, J.; Flint, S.; Brooks, J. Bacterial Cell attachment, the Beginning of a Biofilm. *J. Ind. Microbiol. Biotechnol.* **2007**, *34*, 577–588. [CrossRef]
36. Mah, T.F.C.; O'Toole, G.A. Mechanisms of Biofilm Resistance to Antimicrobial Agents. *Trends Microbiol.* **2001**, *9*, 34–39. [CrossRef]
37. Singh, R.; Ray, P.; Das, A.; Sharma, M. Penetration of Antibiotics through *Staphylococcus aureus* and *Staphylococcus epidermidis* Biofilms. *J. Antimicrob. Chemother.* **2010**, *65*, 1955–1958. [CrossRef] [PubMed]

38. Fagerlund, A.; Langsrud, S.; Heir, E.; Mikkelsen, M.I.; Moretro, T. Biofilm Matrix Composition Affects the Susceptibility of Food Associated Staphylococci to Cleaning and Disinfection Agents. *Front Microbiol.* **2016**, *7*. [CrossRef] [PubMed]
39. Xue, Z.; Lee, W.H.; Coburn, K.M.; Seo, Y. Selective Reactivity of Monochloramine with Extracellular Matrix Components Affects the Disinfection of Biofilm and Detached Cflusters. *Environ. Sci. Technol.* **2014**, *48*, 3832–3839. [CrossRef] [PubMed]
40. Bridier, A.; Dubois-Brissonnet, F.; Greub, G.; Thomas, V.; Briandet, R. Dynamics of the Action of Biocides in *Pseudomonas aeruginosa* biofilms. *Antimicrob. Agents Chemother.* **2011**, *55*, 2648–2654. [CrossRef]
41. Ogunleye, A.; Bhat, A.; Irorere, V.U.; Hill, D.; Williams, C.; Radecka, I. Poly-gamma-glutamic Acid: Production, Properties and Applications. *Microbiol. Sgm* **2015**, *161*, 1–17. [CrossRef] [PubMed]
42. Shemesh, M.; Kolter, R.; Losick, R. The Biocide Chlorine Dioxide Stimulates Biofilm Formation in *Bacillus subtilis* by Activation of the Histidine Kinase KinC. *J. Bacteriol.* **2010**, *192*, 6352–6356. [CrossRef] [PubMed]
43. Strempel, N.; Nusser, M.; Neidig, A.; Brenner-Weiss, G.; Overhage, J. The Oxidative Stress Agent Hypochlorite Stimulates c-di-GMP Synthesis and Biofilm Formation in *Pseudomonas aeruginosa*. *Front. Microbiol.* **2017**, *8*, 2311. [CrossRef] [PubMed]

 *foods*

Article

Biofilm Challenge: Lactic Acid Bacteria Isolated from Bovine Udders versus Staphylococci

Jonathan K. Wallis, Volker Krömker * and Jan-Hendrik Paduch

University of Applied Sciences and Arts Hannover, Faculty II, Department Bioprocess Engineering,
Microbiology, Heisterbergallee 10A, D-30453 Hannover, Germany; jonathan-wallis@hotmail.com (J.K.W.);
Jan-hendrik.paduch@hs-hannover.de (J.-H.P.)
* Correspondence: Volker.Kroemker@hs-hannover.de; Tel.: +49-511-92962205

Received: 26 December 2018; Accepted: 18 February 2019; Published: 20 February 2019

Abstract: Mastitis poses a considerable threat to productivity and to animal welfare on modern dairy farms. However, the common way of antibiotic treatment does not always lead to a cure. Unsuccessful cures can, among other reasons, occur due to biofilm formation of the causative agent. This has attracted interest from researchers to introduce promising alternative therapeutic approaches, such as the use of beneficial lactic acid bacteria (LAB). In fact, using LAB for treating mastitis probably requires the formation of a beneficial biofilm by the probiotic bacteria. The present study investigated the ability of five LAB strains, selected on the basis of results from previous studies, to remove and to replace pathogenic biofilms in vitro. For this purpose, *Staphylococcus (S.) aureus* ATCC 12,600 and two strains—*S. xylosus* (35/07) and *S. epidermidis* (575/08)—belonging to the group of coagulase negative staphylococci (CNS) were allowed to form biofilms in a 96-well plate. Subsequently, the LAB were added to the well. The biofilm challenge was evaluated by scraping off and suspending the biofilm cells, followed by a plate count of serial dilutions using selective media. All the LAB strains successfully removed the staphylococcal biofilms. However, only *Lactobacillus (L.) rhamnosus* ATCC 7469 and *L. plantarum* 2/37 formed biofilms of their own to replace the pathogenic ones.

Keywords: lactic acid bacteria; biofilm; probiotic potential; staphylococci; mastitis

1. Introduction

Bovine mastitis is among the most prevalent and costly diseases the dairy industry is facing today. It has a substantial economic impact as a result of reduced milk yield and poor milk quality, milk losses due to discarded milk after antibiotic treatment, and high costs for drugs and veterinary services [1]. Furthermore, the outcome of antibiotic therapy, which is the common way of treating mastitis, is not always satisfactory [2]. According to Anderl et al. [3], the effect of antimicrobials can be reduced by biofilm formation of the causative agent. Schönborn and Krömker [4] found *Staphylococcus aureus* form biofilms in infected udders. In vitro studies suggest that many more pathogens may cause biofilm-related mastitis [5]. Therefore, novel approaches for treating the disease are needed. Administering probiotic lactic acid bacteria (LAB) is one of the most interesting alternatives to antibiotic treatment and has already shown promising results in previous studies [6–8]. The Food and Agriculture Organization of the United Nations (FAO) defines probiotics as "live microorganisms that, when administered in adequate amounts, confer a health benefit on the host" [9]. Many LAB have been given the GRAS (generally recognized as being safe) status by the Food and Drug Administration (FDA) because they are traditionally used to produce certain foods [9]. Additionally, several members of this group are regarded as commensals of the udder [10] and are therefore presumably harmless to consumers and patients.

According to Frola et al. [11], probiotic bacteria are required to form a beneficial biofilm inside the udder, serving as a barrier against pathogens. The present study investigates the ability of five selected LAB strains to disrupt and replace staphylococcal biofilms with beneficial biofilms of their own in order to exert a probiotic effect.

2. Materials and Methods

2.1. Selection of the Strains

For this study, five LAB strains (Table 1) were selected from the strain collection of the Faculty II, Department for Bioprocess Engineering and Microbiology of the University of Applied Sciences and Arts, Hannover, Germany, according to their biofilm-forming ability and their antimicrobial properties. All the strains had previously shown an ability to inhibit the growth of certain mastitis-causing pathogens [12]. Furthermore, they were all capable of forming a biofilm with a higher-than-average biomass (optical density >0.21 at 570 nm after crystal violet staining) in a recent study [13].

Three staphylococci strains that had already been used in previous studies [12] were selected for the biofilm challenge (Table 1). We chose one *S. aureus* strain as this pathogen is still one of the most important mastitis-causing pathogens and is frequently associated with persistent infections in the udder [14]. The two remaining strains belonged to the coagulase negative staphylococci (CNS) group, a bacterial group of increasing importance in modern dairy herds despite effective mastitis management programs. CNS have been found to cause an increased somatic cell count in infected udder quarters while persisting in the udder for at least 10 months [15]. They are able to induce clinical mastitis in dairy cattle [15]. However, most of the infections caused by CNS remain subclinical [16]. The two CNS strains used in this study (*S. xylosus* (35/07) and *S. epidermidis* (575/08)) were isolated from the udders of cows with mastitis. *S. xylosus* and *S. epidermidis* were among the five most prevalent CNS species isolated from bovine udders in a previous study [17].

Table 1. Bacterial strains used in this study.

Strain	Origin
L. rhamnosus ATCC 7469	American Type Culture Collection
L. plantarum 2/37 *L. brevis* 104/37 *L. plantarum* 118/37	Quarter milk samples with normal secretion (somatic cell count <100,000/mL, no pathogen detected)
L. plantarum 6E	Bedding sample
S. aureus ATCC 12,600	American Type Culture Collection
S. xylosus (35/07) *S. epidermidis* (575/08)	Quarter milk sample from udders of infected cows

2.2. Biofilm Assay

In order to examine the ability of the five LAB strains to disrupt staphylococcal biofilms and to establish probiotic biofilms of their own instead, a method based on Guerrieri et al. [18] was used. First of all, the staphylococci were allowed to preform biofilms. Subsequently, the LAB were added to the staphylococcal biofilms in order to perform the biofilm challenge. The biofilm formation of both species was assessed at three different points in time while incubating the bacteria together.

2.2.1. Preformation of Biofilms by Staphylococci

After transferring the bacteria from the frozen stock culture to the brain heart infusion broth (Carl Roth GmbH+Co. KG, Karlsruhe, Germany), three consecutive subcultures were made, each being incubated aerobically at 37 °C for 24 h. The optical density of the third subculture was then adjusted to 0.6 at 540 nm wavelength corresponding to 7 $\log_{10}$ cfu/mL, and inocula of 200 µL were transferred to the wells of polypropylene 96-well plates (Greiner Bio-One GmbH, Frickenhausen, Germany). Biofilms were grown aerobically at 37 °C for 168 h (seven days). After 72 h, 50% of the broth from each well was replaced by fresh medium. This was performed by removing 100 µL with a pipette. Afterward, the wells were refilled with 100 µL of fresh broth. Then, the plates were incubated for 48 h under the same conditions. Subsequently, 50% of the growth medium was again replaced with fresh brain heart infusion broth, and the 96-well plates were incubated for a further 48 h.

2.2.2. Biofilm Challenge

For the biofilm challenge, LAB inocula were passaged three times, as previously described for the staphylococci. For growing LAB, Tween 80-depleted de Man, Rogosa and Sharpe MRS broth was used, as described by Leccese Terraf et al. [19].

The brain heart infusion broth from the preformed staphylococcal biofilms in the 96-well plates was completely removed with a pipette and replaced with either 200 μL of LAB inoculum or with fresh MRS broth. The wells with fresh MRS broth on preformed staphylococcal biofilms served as negative control. The wells in which LAB were added to the staphylococcal biofilms were the challenge wells. Additionally, for every LAB strain, one well without a preformed pathogenic biofilm was filled with 200 μL inoculum to serve as positive control, and wells without a preformed biofilm were filled with pure MRS broth. The plates were incubated aerobically at 37 °C for 168 h (seven days). Medium refreshment was performed after 72 h, 48 h thereafter, and a further 48 h, as previously described for preformation of the staphylococcal biofilms.

2.2.3. Assessment of Biofilm Formation

Assessment of biofilm formation was carried out along with each medium refreshment for LAB and staphylococci. First, the medium from the wells was discarded and the wells were washed three times with 0.85 % NaCl (*w*/*v*). After that, a sterile cotton wool swab (MWE, Corsham, Wiltshire, UK) was used to scrape off the bacterial cells from the well by pressing the swab against the inner surface and the bottom of the well and rotating it clockwise five times and anti-clockwise a further five times. The cotton tip of the swab was then broken off and dropped into an Eppendorf tube (Eppendorf AG, Hamburg, Germany) containing 1 mL of sterile Ringer's solution (Merck AG, Darmstadt, Germany). This Eppendorf tube was vortexed for 30 s to detach the bacterial cells from the swab. From this suspension, tenfold dilutions were made, and the cfu/mL were determined via plate count using selective media. To detect LAB, MRS agar (Carl Roth GmbH+Co. KG, Karlsruhe, Germany) with a pH value of 5.5 was used to rule out growth of the staphylococci on this medium. Baird Parker agar (Carl Roth GmbH+Co. KG) with 5 % egg yolk tellurite emulsion (Carl Roth GmbH+Co. KG) was used to detect *S. aureus*, and Chapman agar (Carl Roth GmbH+Co. KG) with 5 % egg yolk emulsion (Carl Roth GmbH+Co. KG) was used to detect CNS. Exclusive growth of the bacteria on their specific medium had been confirmed in advance of the assay by performing a plate count from pure overnight cultures.

The whole assay was performed in triplicate.

2.3. Statistical Analysis

Microbial counts (cfu/mL) were converted into logarithmic values. The statistical analysis was performed with IBM SPSS Statistics 24. In order to examine possible effects of the LAB on biofilm growth of staphylococci, the data were subjected to a linear mixed model. Results were regarded as significant when the *p*-value was below 0.05. The staphylococci species and the LAB strains as well as the incubation time served as independent variables. The staphylococci cfu/mL were the dependent variable.

3. Results

3.1. Biofilm Assay

Assessment of Biofilm Formation

All three staphylococci strains showed biofilm formation in the control well. Their biofilms remained detectable until the end of the trial (Figures 1–3). The mean log cfu/mL values from the control wells seemed to decrease over time. *S. aureus* ATCC 12,600 revealed the highest mean cfu/mL values of the three staphylococci, increasing to approximately 7.6 log cfu/mL in the control after 72 h of incubation (Figure 1). In contrast, *S. xylosus* (35/07) showed the lowest staphylococcal cell count, achieving a mean log cfu/mL of approximately 4.4 in the control after 168 h of incubation (Figure 2).

In the wells containing noninoculated MRS broth, we detected no bacteria throughout the trial.

L. rhamnosus ATCC 7469 showed increasing cfu/mL values in the control wells, starting with approximately 5 log mean cfu/mL after 72 h of incubation. After 120 h of incubation, this strain revealed approximately 6 log mean cfu/mL, which remained constant until the end of the trial (Figures 1–3). In the challenge wells containing *L. rhamnosus* ATCC 7469, this was the only detected strain. The mean log cfu/mL values from these wells were similar to those obtained from the control wells (Figures 1–3). We found no biofilm formation by the three investigated staphylococci in the challenge wells after *L. rhamnosus* ATCC 7469 had been added (Figures 1–3).

However, we could still detect biofilm formation by *S. aureus* ATCC 12,600 and *S. xylosus* (35/07) in the challenge well despite the presence of *L. plantarum* 2/37 after 72 h of incubation during one of the three assay repetitions (Figures 1 and 2). Biofilm formation by this strain was neither detected in the challenge nor in the control well after this time span. The first evidence of biofilm formation by *L. plantarum* 2/37 was found after 120 h of incubation in the control wells (approximately 1.3 log cfu/mL) as well as in the challenge wells, where we no longer found biofilms of the three tested pathogens (Figures 1–3). The mean log cfu/mL values in the challenge wells after 120 h of incubation (2.5–4 log cfu/mL) were higher than the values obtained from the controls (Figures 1–3). After 168 h of incubation, *L. plantarum* 2/37 maintained a biofilm in the challenge wells against all the three investigated staphylococci and was still present in the control wells, with the calculated values in both kinds of wells being more similar (4.1–4.8 mean log cfu/mL).

In the challenge wells containing the strains *L. brevis* 104/37, *L. plantarum* 118/37, and *L. plantarum* 6E, no staphylococcal biofilms were found after 72 h of incubation. However, none of them formed a detectable biofilm of their own either in the control or in the challenge well.

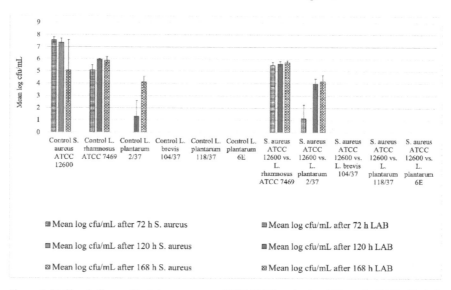

Figure 1. Biofilm challenge: *Staphylococcus aureus* ATCC 12,600 vs. lactic acid bacteria (LAB). Cfu/mL values are shown transformed by log (± standard error of the mean).

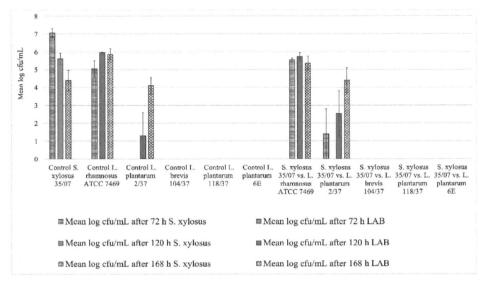

Figure 2. Biofilm challenge: *S. xylosus* (35/07) vs. LAB. Cfu/mL values are shown transformed by log (± standard error of the mean).

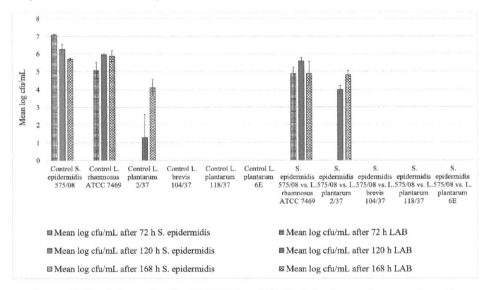

Figure 3. Biofilm challenge: *S. epidermidis* (575/08) vs. LAB. Cfu/mL values are shown transformed by log (± standard error of the mean).

3.2. Statistical Analysis

The statistical analysis revealed a significant reduction in staphylococcal growth by LAB ($p < 0.05$). Furthermore, the incubation time significantly affected the reduction ($p < 0.05$). We observed no differences between the five investigated LAB strains.

4. Discussion

The method for evaluating biofilm formation applied in this study represents a culture-based approach involving specific growth media in order to differentiate between LAB and staphylococci. According to Jahid and Ha [20], culture-based methods are the most useful technique to differentiate known strains from mixed-species biofilms. The successful use of Baird Parker and MRS agars to distinguish *S. aureus* and LAB populations was already described by Gonzalez et al. [21]. The crystal violet assay is a common method used to assess biofilm formation. Crystal violet binds nonspecifically to viable and to dead bacterial cells as well as to matrix components [22]. Therefore, measuring the optical density after crystal violet staining is a valuable tool to establish the total biomass of a biofilm. However, it cannot distinguish between different species in a mixed-species biofilm. For this reason, crystal violet staining was not an option for evaluating the specific share in a biofilm of staphylococci and LAB. Nevertheless, the LAB strains included in this study were selected on the basis of the results of a crystal violet assay performed in a previous study [13], where a strong biofilm had formed on a polypropylene surface after 72 h incubation under the same conditions provided in this study. Therefore, we assume that it was due to the culture-based method for biofilm quantification that we found no biofilm formation by three of the LAB strains and not due to the growth conditions. Fernández Ramírez et al. [22] stated that results of a crystal violet assay might correlate poorly with those obtained by culture-based methods as not all the stained biomass in a mature biofilm has to consist of culturable bacterial cells. These findings could explain why we did not observe biofilm formation by *L. brevis* 104/37, *L. plantarum* 118/37, and *L. plantarum* 6E in the present study. Furthermore, Klinger-Strobel et al. [23] stated that loss of biomass could occur due to the washing step commonly performed in crystal violet assays. However, this might account even more for the culture-based technique applied in our study as the biofilms were washed three times using pipette suction in order to remove unbound cells prior to scraping off the biofilm by rotating a cotton swab in the well. The previously performed crystal violet assay involved only one washing step using gently flowing tap water.

We could not find any biofilm formation by staphylococci in the challenge well after LAB had been added to it, except for one repetition of the assay during which we detected *S. aureus* ATCC 12,600 and *S. xylosus* (35/07) biofilms in the first assessment of the challenge against *L. plantarum* 2/37. As we found no evidence of LAB and staphylococci being present in the same well at the same time, we can deduce that there was no formation of mixed-species biofilms containing both LAB and staphylococci. The staphylococci maintained a strong biofilm in the control well containing MRS broth where LAB were absent. Therefore, we assume that the LAB were responsible for eradicating the staphylococcal biofilm from the challenge well, and the effect was not due to the MRS broth. Furthermore, the statistical analysis revealed a significant growth reduction ($p < 0.05$).

L. rhamnosus ATCC 7469 appeared to be very effective at removing biofilms formed by staphylococci. It might be suitable for a probiotic remedy due to its high growth rates and its ability to form a strong biofilm after a short period of time. According to James et al. [24], high growth rates may lead to dominance over other biofilm formers when existing in the same habitat. Nonetheless, this strain showed a below-average adhesion to epithelial cells from the bovine udder in previous in vitro studies [13], which might interfere with the strain's ability to form a beneficial biofilm in the udder under in vivo conditions. *L. plantarum* 2/37 seems to be a rather slow-growing strain. As adhesion to the epithelium and subsequent biofilm formation accounts for the ability of a potential probiotic strain to maintain its presence in the host and its positive effects over time [25], slow formation of a stable biofilm might be a disadvantage. However, *L. plantarum* 2/37 did finally form a stable biofilm and showed a strong adhesion ability to epithelial cells of the bovine udder during previous investigations [13]. Therefore, this strain might still be a potential candidate for a probiotic remedy. *L. brevis* 104/37, *L. plantarum* 118/37, and *L. plantarum* 6E revealed the ability to eradicate staphylococcal biofilms fast and effectively. Nonetheless, these strains were neither able to form a detectable biofilm of their own in the control nor in the challenge well. The three aforementioned strains showed a strong antimicrobial activity, which is in line with the results of Diepers et al. [12]. However, their inability to form a detectable biofilm of their own might interfere with their probiotic potential, as previously explained for *L. plantarum* 2/37.

With regard to mastitis treatment based on LAB, further research is needed, including in vivo studies, as the bacteria might show a different behavior concerning biofilm formation in a milky environment [26]. Additionally, their safety for consumers and patients is yet to be verified, since mastitis by LAB as well as severe infections in humans are described in literature [9,27].

5. Conclusions

The present study focused on the ability of five LAB strains to disrupt and replace pathogenic biofilms formed by staphylococci with a presumably beneficial biofilm of their own in vitro. The results recommend two strains—*L. rhamnosus* ATCC 7469 and *L. plantarum* 2/37—for further investigations, focusing on their safety for consumers and patients as well as their beneficial properties on udder health under in vivo conditions.

Author Contributions: Conceptualization, J.K.W., V.K., and J.-H.P.; methodology, J.K.W.; validation, J.K.W.; formal analysis, V.K.; investigations, J.K.W.; resources, V.K. and J.-H.P.; writing—original draft preparation, J.K.W.; writing—review and editing, J.K.W., V.K., and J.-H.P.; visualization, J.K.W.; supervision, J.-H.P.; project administration, V.K.; funding acquisition, V.K. and J.-H.P.

Funding: The Steinbeis Research Center Milk Science and DBU (Deutsche Bundesstiftung Umwelt; Project 31833) provided financial support for our research.

Conflicts of Interest: The authors declare no conflict of interest. The funders had no role in the design of the study, in the collection, analyses, or interpretation of data, nor the writing of the manuscript, or in the decision to publish the results.

References

1. Halasa, T.; Huijps, K.; Østerås, O.; Hogeveen, H. Economic effects of bovine mastitis and mastitis management: A review. *Vet Q.* **2007**, *29*, 18–31. [CrossRef] [PubMed]
2. Linder, M.; Paduch, J.H.; Grieger, A.S.; Mansion-de Vries, E.; Knorr, N.; Zinke, C.; Teich, K.; Krömker, V. Cure rates of chronic subclinical *Staphylococcus aureus* mastitis in lactating dairy cows after antibiotic therapy./Heilungsraten chronischer subklinischer Staphylococcus aureus-Mastitiden nach antibiotischer Therapie bei laktierenden Milchkühen. *Berl. Münchener Tierärztliche Wochenschr.* **2013**, *126*, 291–296.
3. Anderl, J.N.; Franklin, M.J.; Stewart, P.S. Role of antibiotic penetration limitation in *Klebsiella pneumoniae* biofilm resistance to ampicillin and ciprofloxacin. *Antimicrob. Agents Chemother.* **2000**, *44*, 1818–1824. [CrossRef] [PubMed]
4. Schönborn, S.; Krömker, V. Detection of the biofilm component polysaccharide intercellular adhesin in *Staphylococcus aureus* infected cow udders. *Vet Microbiol.* **2016**, *196*, 126–128. [CrossRef] [PubMed]
5. Schönborn, S.; Wente, N.; Paduch, J.H.; Krömker, V. In vitro ability of mastitis causing pathogens to form biofilms. *J. Dairy Res.* **2017**, *84*, 198–201. [CrossRef] [PubMed]
6. Beecher, C.; Daly, M.; Berry, D.P.; Klostermann, K.; Flynn, J.; Meaney, W.; Hill, C.; Mc Carthy, T.V.; Ross, R.P.; Giblin, L. Administration of a live culture of *Lactococcus lactis* DPC 3147 into the bovine mammary gland stimulates the local host immune response, particularly IL-1β and IL-8 gene expression. *J. Dairy Res.* **2009**, *76*, 340. [CrossRef] [PubMed]
7. Crispie, F.; Alonso-Gomez, M.; O'Loughlin, C.; Klostermann, K.; Flynn, J.; Arkins, S.; Meaney, W.; Ross, R.P.; Hill, C. Intramammary infusion of a live culture for treatment of bovine mastitis: Effect of live lactococci on the mammary immune response. *J. Dairy Res.* **2008**, *75*, 374–384. [CrossRef] [PubMed]
8. Klostermann, K.; Crispie, F.; Flynn, J.; Ross, R.P.; Hill, C.; Meaney, W. Intramammary infusion of a live culture of *Lactococcus lactis* for treatment of bovine mastitis: Comparison with antibiotic treatment in field trials. *J. Dairy Res.* **2008**, *75*, 365–373. [CrossRef]
9. FAO. Probiotics in animal nutrition: Production, impact and regulation. In *Animal Production and Health Paper*; Food and Agriculture Organization of the United Nations (FAO): Rome, Italy, 2016.

10. Hagi, T.; Sasaki, K.; Aso, H.; Nomura, M. Adhesive properties of predominant bacteria in raw cow's milk to bovine mammary gland epithelial cells. *Folia Microbiol.* **2013**, *58*, 515–522. [CrossRef]
11. Frola, I.D.; Pellegrino, M.S.; Espeche, M.C.; Giraudo, J.A.; Nader-Macias, M.E.F.; Bogni, C.I. Effects of intramammary inoculation of *Lactobacillus perolens* CRL1724 in lactating cows' udders. *J. Dairy Res.* **2012**, *79*, 84–92. [CrossRef]
12. Diepers, A.-C.; Krömker, V.; Zinke, C.; Wente, N.; Pan, L.; Paulsen, K.; Paduch, J.H. In vitro ability of lactic acid bacteria to inhibit mastitis-causing pathogens. *Sustain. Chem. Pharm.* **2017**, *5*, 84–92. [CrossRef]
13. Wallis, J.K.; Krömker, V.; Paduch, J.H. Biofilm formation and adhesion to bovine udder epithelium of potentially probiotic lactic acid bacteria. *AIMS Microbiol.* **2018**, *4*, 209–224. [CrossRef]
14. Artursson, K.; Soderlund, R.; Liu, L.; Monecke, S.; Schelin, J. Genotyping of Staphylococcus aureus in bovine mastitis and correlation to phenotypic characteristics. *Vet Microbiol.* **2016**, *193*, 156–161. [CrossRef] [PubMed]
15. Gillespie, B.E.; Headrick, S.I.; Boonyayatra, S.; Oliver, S.P. Prevalence and persistence of coagulase-negative Staphylococcus species in three dairy research herds. *Vet Microbiol.* **2009**, *134*, 65–72. [CrossRef] [PubMed]
16. Krömker, V. *Kurzes Lehrbuch Milchkunde und Milchhygiene*, 1st ed.; Parey: Stuttgart, Germany, 2007; p. 60.
17. Fry, P.R.; Middleton, J.R.; Dufour, S.; Perry, J.; Scholl, D.; Dohoo, I. Association of coagulase-negative staphylococcal species, mammary quarter milk somatic cell count, and persistence of intramammary infection in dairy cattle. *J. Dairy Sci.* **2014**, *97*, 4876–4885. [CrossRef] [PubMed]
18. Guerrieri, E.; de Niederhausern, S.; Messi, P.; Sabia, C.; Iseppi, R.; Anacarso, I.; Bondi, M. Use of lactic acid bacteria (LAB) biofilms for the control of *Listeria monocytogenes* in a small-scale model. *Food Control.* **2009**, *20*, 861–865. [CrossRef]
19. Leccese Terraf, M.C.; Juárez Tomás, M.S.; Nader-Macías, M.E.F.; Silva, C. Screening of biofilm formation by beneficial vaginal lactobacilli and influence of culture media components. *J. Appl. Microbiol.* **2012**, *113*, 1517–1529. [CrossRef] [PubMed]
20. Jahid, I.K.; Ha, S.D. The Paradox of Mixed-Species Biofilms in the Context of Food Safety. *Compr. Rev. Food Sci. Food Saf.* **2014**, *13*, 990–1011. [CrossRef]
21. Gonzalez, S.; Fernandez, L.; Campelo, A.B.; Gutierrez, D.; Martinez, B.; Rodriguez, A.; García, P. The Behavior of *Staphylococcus aureus* Dual-Species Biofilms Treated with Bacteriophage phiIPLA-RODI Depends on the Accompanying Microorganism. *Appl. Environ. Microbiol.* **2017**, *83*, 14. [CrossRef]
22. Fernández Ramírez, M.D.; Smid, E.J.; Abee, T.; Nierop Groot, M.N. Characterisation of biofilms formed by *Lactobacillus plantarum* WCFS1 and food spoilage isolates. *Int. J. Food Microbiol.* **2015**, *207*, 23–29. [CrossRef]
23. Klinger-Strobel, M.; Suesse, H.; Fischer, D.; Pletz, M.W.; Makarewicz, O. A Novel Computerized Cell Count Algorithm for Biofilm Analysis. *PLoS ONE* **2016**, *11*, e0154937. [CrossRef] [PubMed]
24. James, G.A.; Beaudette, L.; Costerton, J.W. Interspecies bacterial interactions in biofilms. *J. Indust. Microbiol.* **1995**, *15*, 257–262. [CrossRef]
25. Tallon, R.; Arias, S.; Bressollier, P.; Urdaci, M.C. Strain- and matrix-dependent adhesion of *Lactobacillus plantarum* is mediated by proteinaceous bacterial compounds. *J. Appl. Microbiol.* **2007**, *102*, 442–451. [CrossRef] [PubMed]
26. Varhimo, E.; Varmanen, P.; Fallarero, A.; Skogman, M.; Pyorala, S.; Iivanainen, A.; Sukura, A.; Vuorela, P.; Savijoki, K. Alpha- and b-casein components of host milk induce biofilm formation in the mastitis bacterium *Streptococcus uberis*. *Vet Microbiol.* **2011**, *149*, 381–389. [CrossRef] [PubMed]
27. Plumed-Ferrer, C.; Uusikyla, K.; Korhonen, J.; von Wright, A. Characterization of *Lactococcus lactis* isolates from bovine mastitis. *Vet Microbiol.* **2013**, *167*, 592–599. [CrossRef] [PubMed]

Article

Antibiofilm Effect of DNase against Single and Mixed Species Biofilm

Komal Sharma [1] and Ankita Pagedar Singh [2,*]

[1] Ashok & Rita Patel Institute of Integrated Study and Research in Biotechnology and Allied Sciences,
    New Vallabh Vidya Nagar, Anand 388121, Gujarat, India; komalsharma0325@gmail.com
[2] Department of Food Processing Technology, AD Patel Institute of Technology, New Vallabh Vidya Nagar,
    Anand 388121, Gujarat, India
* Correspondence: ankitapagedar@gmail.com; Tel.: +91-269-223-3680

Received: 30 November 2017; Accepted: 15 March 2018; Published: 19 March 2018

Abstract: Biofilms are aggregates of microorganisms that coexist in socially coordinated micro-niche in a self-produced polymeric matrix on pre-conditioned surfaces. The biofilm matrix reduces the efficacy of antibiofilm strategies. DNase degrades the extracellular DNA (e-DNA) present in the matrix, rendering the matrix weak and susceptible to antimicrobials. In the current study, the effect of DNase I was evaluated during biofilm formation (pre-treatment), on preformed biofilms (post-treatment) and both (dual treatment). The DNase I pre-treatment was optimized for *P. aeruginosa* PAO1 (model biofilm organism) at 10 µg/mL and post-treatment at 10 µg/mL with 15 min of contact duration. Inclusion of Mg^{2+} alongside DNase I post-treatment resulted in 90% reduction in biofilm within only 5 min of contact time (irrespective of age of biofilm). On extension of these findings, DNase I was found to be less effective against mixed species biofilm than individual biofilms. DNase I can be used as potent antibiofilm agent and with further optimization can be effectively used for biofilm prevention and reduction in situ.

Keywords: biofilms; DNase I; pre-treatment; post-treatment; mixed species biofilm; disintegration of matrix; antibiofilm methods

---

1. Introduction

Microorganisms prefer to coexist in an extremely coordinated surface adhered lifestyle, known as biofilm. Biofilms are a grave concern across various industries like food, textile, paper, oil, aviation, shipping and even the medical sector. They have accounted for reduced efficacy of heat exchange processes, corrosion of materials, blocking of membranes and degradation of ship hulls. Biofilms mediated infections contribute to almost 80% of clinical infections reported globally [1]. In the food industry, biofilms on food contact surfaces pose a food safety hazard and product quality issues. Antibiofilm strategies in the healthcare sector include use of antibiotics and/or biocides. However, in food industry scrapping, hot water treatment, acid/alkali treatments and biocides as a part of the cleaning regime are used to combat biofilms. As biofilms are notorious for being resistant to conventional antibiofilm approaches, alternative antibiofilm strategies like using proteases, amylases, bis-(3′-5′)-cyclic dimeric guanosine monophosphate (c-di-GMP) and quorum sensing inhibitors have been explored [2]. These methods are reportedly more effective for prevention of biofilm formation and may or may not be effective on pre-formed biofilms [3–5].

For development of an effective antibiofilm strategy, a thorough understanding of the biofilm formation process (initial adhesion, maturation, quorum sensing and dissemination), a metabolic state of biofilm inhabitants and composition of biofilm matrix is required. Biofilm inhabitants display a reduced metabolic rate, enhanced efflux, adaptive and cross-resistances and hence are more resistant to antimicrobials than their planktonic counterparts [6,7]. In addition, the biofilm matrix acts as a protective barrier and reduces the percolation of antimicrobials to a deeper strata of biofilm structure. The biofilm matrix is composed of 40–95% polysaccharides, 1–60% proteins, 1–40% lipids and 1–10% nucleic acid [8]. Prevalence of the e-DNA in the biofilm matrix has been reported in biofilms of several microorganisms, both Gram-positive and Gram-negative. The release of e-DNA is mediated by autolysis (programmed cell death i.e., suicide and altruistic cell death-fratricide) [9] and through vacuoles and membrane vesicles [10]. The e-DNA contributes to cell-surface and cell–cell interactions, horizontal gene transfer, integrity, cohesivity and viscoelasticity of the biofilm matrix and thus plays a major role in biofilm stability [11].

In view of significance of e-DNA in biofilm matrix and biofilm formation [11], it is indeed a potential target for development of antibiofilm strategies by using DNA degrading enzymes i.e., DNase. Antibiofilm effect of DNase has been studied for organisms such as *S. aureus* and *P. aeruginosa*, *E. coli*, *Acinetobacter baumannii*, *Haemophilus influenzae* and *K. pneumoniae*. Most of these studies have been conducted using commercially available DNases like DNase I (derived from bovine pancreas), DNase 1L2 (human keratinocyte DNase), Dornase alpha (recombinant human DNase), λ exonuclease (viral DNase), NucB and streptodornase produced by *Bacillus licheniformis* and *Streptococcus* spp., respectively [12]. Moreover, microorganisms producing nucleases have been shown to form lesser biofilm than their non-nuclease producing mutants [13]. Addition of L-methionine that induces DNase secretion by *P. aeruginosa*, resulted in reduced biofilm formation [14]. The antibiofilm effect of DNase has been studied with or without antibiotics, dispersinB [15,16] and glutathione [17]. Previous studies have reported that, in the presence of DNase, a lower concentration of antibiotics was required to inhibit biofilm formation by *Campylobacter jejuni* [18,19]. Most of the published studies have used DNase in growth medium itself i.e., its addition at time point 0 of biofilm formation. In the current investigation, such a biofilm preventive effect has been described using the term "pre-treatment".

On the other hand, DNase based treatments of pre-formed biofilms have not been explored much. There are only a few reports available [20,21], which describe application of DNase in combination with proteinase K [22,23], EDTA [24] and dextranase [25] on already formed biofilms. Such a biofilm control/therapeutic effect has been discussed using the term "Post-treatment" in this study. Most of these studies have been carried out on single species biofilm and thus mixed species biofilms yet remain to be explored. Moreover, the antibiofilm effect of DNase evaluated on a single organism may not be directly applicable on in situ biofilms that are formed by mixed species consortia. In purview of reviewed literature, it appears that a study comparing the effect of DNase on biofilms formed by different pathogens will be an addition to the existing knowledge. Therefore, in the current investigation, the antibiofilm efficacy of DNase I treatments (pre and post) were optimized on *P. aeruginosa* PAO1 biofilms. *Pseudomonas* spp. is a concern in food industry due to its inherent antimicrobial resistance and potential to produce heat stable proteases and lipases [26]. Though *P. aeruginosa* is not a typical food related pathogenic organism, its presence in drinking water poses a health hazard [27]. In addition, *P. aeruginosa* forms copious biofilm and thus is considered as a model organism for biofilm formation. In this study, the DNase I treatments optimized using *P. aeruginosa* PAO1 were extended to mixed-species biofilm of organisms (*Staphylococcus aureus*, *Klebsiella* spp., *Enterococcus faecalis*, *Salmonella* Typhimurium) that are relevant to food industry.

2. Materials and Methods

2.1. Culture Maintenance

Microorganisms used in the current investigation were *Pseudomonas aeruginosa* PAO1 (MTCC 3541), *Enterococcus faecalis* (ATCC 29212), *Salmonella* Typhimurium (ATCC 23564) and *Staphylococcus aureus* (ATCC 25923). These cultures were obtained either from the Microbial Type Culture Collection (MTCC) at the Institute of Microbial Technology, Chandigarh, India or American Type Culture Collection (ATCC), Manassas, VA, USA. One *Klebsiella* spp. that was isolated from a biofilm sample obtained from the food industry was also used in this study. The cultures were maintained in tryptone soy broth

(TSB) or on agar plates and stored as glycerol stocks at -40 °C. Culture inoculum for experiments was prepared by adjusting optical density (OD) of overnight activated culture to 0.5 (c.a., 8 Log cfu/mL) at 620 nm. All materials and reagents were procured from HiMedia Labs, Mumbai, India, unless specified otherwise.

2.2. Biofilm Formation Assay

The assay was carried out in accordance with a previously published protocol [28]. Briefly, 200 µL of TSB per well of sterile 96-well plate made of polystyrene (Axiva Biotech, New Delhi, India) was inoculated with 20 µL of inoculum and the biofilm was allowed to develop at 37 °C/24 h. Later, the contents of the wells were decanted and wells were washed 3–4 times with sterile PBS (Phosphate Buffered Saline) to dislodge the loosely adhered cells. The remaining biofilms were vigorously blotted on stack of paper towels and air dried [29]. The biofilms were stained with 1% crystal violet, rinsed 3–4 times with water in a large Petri dish to remove the excess stain, blotted on stack of paper towels and air-dried. The crystal violet bound to biofilms was then resolubilized using 33% glacial acetic acid and absorbance was measured at 595 nm (A595 nm; plotted on the primary *y*-axis) using a microplate reader (EPOCH 2c, BIOTEK, Winooski, VT, USA). As a negative control, uninoculated wells containing TSB were treated similarly and readings obtained were subtracted from the test readings.

2.3. Optimization of DNase I Concentration for Pre-Treatment

A gradient of DNase I in the range of 0–50 µg/mL was prepared by dissolving lyophilized powder in nuclease free water and diluting it with 0.15 M NaCl solution to achieve the desired concentrations. For optimization of DNase I concentration for pre-treatment, biofilms of *P. aeruginosa* PAO1 were formed in the presence of different concentrations of DNase I for 24, 48, 72, 96 h and quantified as A595 nm. As negative control, uninoculated wells containing TSB and diluent were treated similarly and readings obtained were subtracted from the test readings. Positive control wells containing inoculated TSB without DNase I were considered as "Control A595 nm". The biofilm percentage reduction (BPR was calculated as below and plotted on the secondary *y*-axis:

$$BPR = \left(\frac{Control\ A595\ nm - test\ A595\ nm}{Control\ A595\ nm} \right) \times 100$$

2.4. Optimization of Contact Time and Concentration of DNase I for Post Treatment

As described in previous sections, biofilms of *P. aeruginosa* PAO1 were formed for 24, 48, 72, 96 h. The contents of the plate were decanted and rinsed using sterile PBS. The wells containing pre-formed biofilms were refilled with TSB containing DNase I at a concentration optimized for pre-treatment. The plate was left undisturbed to maintain contact duration of 0, 5, 10, 15, 20, 25, 30, 35, 60, 75 and 120 min. Subsequently, the plate was decanted, rinsed, air-dried, stained, destained and quantified as A595 nm. Test controls containing only diluents (0.15 M NaCl) were also evaluated for antibiofilm effects, if any, for respective contact times.

Furthermore, the concentration of DNase I was also optimized in the presence of Mg^{2+} ions (10 mM) for a contact time of 15 min. The antibiofilm effect of Mg^{2+}, if any, was also evaluated by setting up a test control containing only Mg^{2+} in absence of DNase I.

2.5. Pre-Treatment, Post-Treatment and Dual Treatment of Microbial Biofilms by DNase I

Individual biofilm formation by *P. aeruginosa* PAO1, *E. faecalis*, *S.* Typhimurium, *S. aureus* and *Klebsiella* spp. was done in TSB for 24 h and subjected to pre-treatment, post-treatment and dual treatment using DNase I (without Mg^{2+}) as described earlier. Biofilm quantification was done in terms of A595 nm.

2.6. Pre-Treatment, Post-Treatment and Dual Treatment of Mixed Species Biofilm by DNase I

In order to prepare mixed species consortium, either of the pathogen was added in 2× concentration than others in a cocktail (for example: in *P. aeruginosa* PAO1 2× cocktail, the ratio of test pathogens *P. aeruginosa* PAO1: *S. aureus*: *Salmonella* Typhimurium: *E. faecalis*: *Klebsiella* spp. was 2:1:1:1:1). Similarly, 2× cocktails with one of the pathogens as dominant were also prepared, namely, *S. aureus* 2×, *Salmonella* 2×, *E. faecalis* 2×, *Klebsiella* spp. 2× and used for biofilm formation for 24 and 48 h. The biofilms were subjected to DNase I pre-treatment, post-treatment and dual treatment (without Mg^{2+}) and quantified as described in previous sections.

2.7. Statistical Analysis

All the experiments were conducted in triplicate and minimum three trials were carried out for each experiment. The results were calculated as average values of three readings along with standard deviation depicted as error bars. The average, standard deviation, for the readings obtained was determined by using Microsoft Excel Software (Microsoft Office 2010, Redmond, WA, USA). Statistical tool XL-statistics v4.5 was used for carrying out Student's *t*-test (with Bonferroni post hoc analysis) and Analysis of variance (ANOVA). The statistical tool is a freeware of set of workbooks for Microsoft excel and available online [29]. 'Significance' is expressed at the 5% level ($p < 0.05$) or mentioned otherwise.

3. Results

3.1. Optimization of Pre-Treatment and Post-Treatment

3.1.1. DNase I Concentration for Pre-Treatment

P. aeruginosa PAO1 biofilm was developed in varying concentrations of DNase I (0–50 μg/mL; without Mg^{2+}) for 24, 48, 72, and 96 h. The results are expressed in terms of both biofilm quantification (A595 nm; Figure 1) and biofilm percentage reduction (BPR) (Figure 1). In comparison to control biofilm (DNase concentration 0 μg/mL), a reduction of 68.6% was observed when biofilm was grown for 24 h in the presence of 5 μg/mL of DNase I. The BPR observed for biofilms cultivated for 48, 72, 96 h in 5 μg/mL of DNase I was only 36%, 10%, 7%, respectively. On the other hand, when biofilms were cultivated in the presence of 10 μg/mL DNase I, BPR was found to be 70%, 50%, 48%, 26% for 24, 48, 72, 96 h old biofilms, respectively. Further increase in concentration of DNase I, beyond 10 μg/mL, did not result in significant difference in BPR ($p > 0.05$). The susceptibility of 96 h biofilm was least at all the DNase I concentrations tested. However, the susceptibility of biofilms when cultivated for 48 and 72 h was almost at par (50% biofilm reduction). Based on these findings, 10 μg/mL of DNase I was selected as the optimal concentration for pre-treatment.

3.1.2. DNase I Contact Time for Post-Treatment

The preformed biofilms of *P. aeruginosa* PAO1 biofilm were treated with 10 μg/mL of DNase I (without Mg^{2+}) for varying contact duration ranging from 0 to 120 min. However, the result as presented in Figure 2a has been shown only until 35 min of contact duration as the observations at other contact durations were more or less similar. Irrespective of the age of biofilm, BPR observed was in the range of 45–53% at contact duration of 5 min and 73–77% for contact duration of 10 min of DNase I treatment. Notably, insignificant difference, irrespective of the age of the biofilms, was observed in BPR when post-treatment was done for more than 10 min ($p > 0.05$; Figure 2a). Thus, to be on the safer side, 15 min of contact duration was selected for post-treatment.

Further efficacy of DNase I for post treatment was evaluated in the presence of Mg^{2+} ions (10 mM). It was found that, in the presence of Mg^{2+}, DNase I could effectively reduce *P. aeruginosa* PAO1 biofilm by 90% at a concentration of 5 μg/mL irrespective of the age of biofilm (Figure 2b). Increasing the concentration of DNase I in the presence of Mg^{2+} did not result in significant difference in biofilm reduction ($p > 0.05$). A control containing only Mg^{2+} (without DNase I) did not exert any biofilm reduction effect.

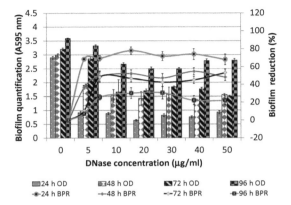

Figure 1. Effect of DNase I (without Mg^{2+}) pretreatment on *P. aeruginosa* PAO1 biofilm grown for 24, 48, 72, 96 h in varying concentrations of DNase I (0–50 μg/mL). Biofilm quantification (A595 nm) on the primary *y*-axis and biofilm percentage reduction (line graph) on the secondary *y*-axis. OD: optical density; BPR: biofilm percentage reduction.

In view of these observations, DNase I concentration in the presence of Mg^{2+} was optimized over a range of 0 to 5 μg/mL in steps of 0.5 while keeping the contact duration constant at 15 min. In this experiment, in addition to polystyrene, antibiofilm efficacy of DNase I was also evaluated on polypropylene. It was found that 1.5 μg/mL and 2 μg/mL of DNase I could effectively reduce the 24 h old *P. aeruginosa* PAO1 biofilm by 80% on polystyrene and 75% on polypropylene, respectively (Figure 2c). The same assay was reconducted at constant DNase I concentration (1.5 μg/mL for Polystyrene and 2 μg/mL for polypropylene), but, for variable contact time, showed that aforementioned antibiofilm efficacy could be achieved within only 5 min of contact duration. It is apparent from the above-mentioned results that Mg^{2+} ions are essential for antibiofilm efficacy of DNase I.

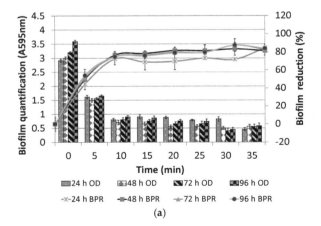

(a)

Figure 2. *Cont.*

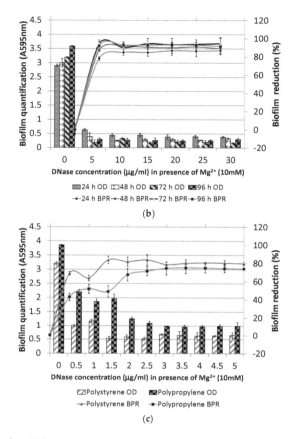

Figure 2. Effect of variable time of post-treatment on 24, 48, 72, 96 h old *P. aeruginosa* PAO1 biofilm with (**a**) DNase (10 μg/mL; without Mg^{2+}) and for varying contact time; (**b**) DNase (10 μg/mL) in the presence of Mg^{2+} (10 mM) and for contact time of 15 min; (**c**) variable DNase (μg/mL) in the presence of Mg^{2+} (10 mM) and for contact time of 15 min.

3.2. DNase I Treatment (Pre, Post and Dual) of Individual and Mixed Species Biofilms

The effect of DNase I treatments (without Mg^{2+}) on biofilm formation was evaluated in three sets of experiments: (Case-A) Individual pathogen: One test pathogen alone was used as inoculum to form biofilm for 24 h (Figure 3a); (Case-B) Pathogen 2× 24 h: mixed biofilm was formed for 24 h using all the pathogens at 1× inoculum level except one that was used at the 2× inoculum level. Hence, five different biofilms that were initiated with inoculum having one organism out of five at the 2× level and remaining at 1× (Figure 3b); (Case-C) Pathogen 2× 48 h: Similar to case b except the age of biofilm, which was 48 h (Figure 3c).

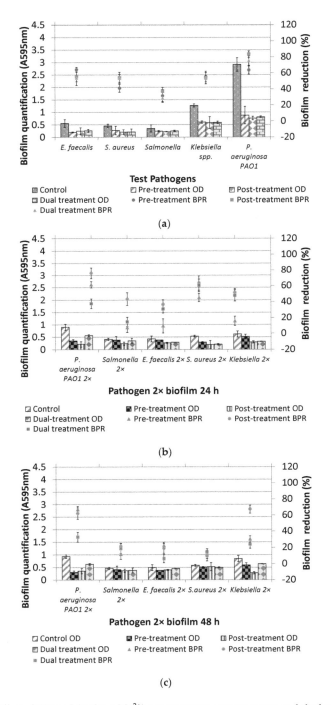

Figure 3. Effect of DNase I (without Mg^{2+}) pre-treatment, post-treatment and dual treatment on biofilms formed by (**a**) test organism (individual); (**b**) test organism 2× mixed species biofilm formed for 24 h; (**c**) test organism 2× mixed species for biofilm formed for 48 h.

The BPR as a result of three treatments (Pre, Post and Dual) seems to be similar for case-A, as evident by overlapping error bars in Figure 3a and statistical insignificant difference ($p > 0.05$). In reference to case-B, overall post-treatment was significantly better than pre-treatment ($p < 0.05$) but on par with dual-treatment ($p > 0.05$) except *P. aeruginosa* PAO1 2×, *Salmonella* Typhimurium 2×. Pre-treatment of *Salmonella* Typhimurium 2×, *E. faecalis* 2×, *Klebsiella* 2× with DNase I resulted in BPR of 7%, 9% and 15%, respectively. It is interesting to note that using pathogen at 2× inoculum level did not result in greater biofilm formation than when they were used at 1× inoculum level, not even in the case of *P. aeruginosa* PAO1 2×. However, at extended incubation time, control biofilm in case-C was significantly greater than that of case B ($p < 0.05$). In terms of BPR data, the efficacy of post-treatment was found to be reduced ($p > 0.05$) when pathogen 2× biofilm was grown for 48 h.

4. Discussion

Owing to several roles played by e-DNA in biofilm formation and strengthening of biofilm matrix, it has recently received much deserved attention by the research community. The current study involves pathogens like model biofilm forming organism *P. aeruginosa* PAO1 and other test organisms viz. *Klebsiella* spp., *S. aureus*, *E. faecalis* and *Salmonella* Typhimurium. These organisms either display biofilm mediated pathogenesis, or form enhanced biofilm in the presence of e-DNA, or are of relevance to the food industry [30]. The effect of DNase I on biofilm formation by test organisms was evaluated by pre-treatment, post-treatment and dual treatment. Optimization of treatments were done using *P. aeruginosa* PAO1 and the effect of the optimized treatments was evaluated on biofilm formation potential of individual and 24 and 48 h old mixed species biofilm.

The findings indicate that DNase I pretreatment (10 µg/mL) resulted in BPR of 68%. These findings can be corroborated with a previously published study that reported 40% reduction in biofilms when grown in the presence of 5 µg/mL DNase for 24 h [31]. Pretreatment of DNase at a concentration 5 µg/mL has been reported to reduce *E. coli* and *S. aureus* biofilm by 47–54% [32]. It can be concluded that DNase I concentration optimized in the current study for pre-treatment of biofilms is therefore comparable to the published literature.

Interestingly, we observed an inverse relation between the antibiofilm effect of DNase I pre-treatment and age of biofilm. Reduced vulnerability of the aged biofilm to DNase I indicates lower dependence of such biofilms on e-DNA. The mature biofilms might also scavenge the e-DNA in biofilm matrix to use it as a source of nutrition [8]. Moreover, the presence of DNase I throughout the process of biofilm formation (as during pre-treatment) may have propelled the biofilm to devise alternative strategies to compensate for the roles e-DNA plays [33]. The mature biofilm may still have e-DNA in the matrix but strengthening of the matrix by methods other than e-DNA in mature biofilms might render DNase ineffective. The results obtained in the current study are in complete agreement with a very recently published study wherein the antibiofilm effect of DNase was reported to be diminishing with the advancing age of biofilm [34]. Moreover, the efficacy of DNase is also dependent on availability of Mg^{2+} ion as discussed in detail in the following section.

After optimization of pre-treatment, post treatment was optimized at 10 µg/mL DNase I and contact duration of 15 min that resulted in 73–77% BPR. Most of the published studies have reported lower BPR at higher DNase concentrations and contact duration. A previously published study has reported 50% reduction in clinical *P. aeruginosa* biofilms when post-treated with glutathione and DNase (40U) [17]. Using DNase concentration almost 200 times higher, only 60% BPR could be achieved against preformed biofilms of *Acinetobacter baumannii* [10]. In another study with *Helicobacter pylori*, a very high concentration of DNase (1000 µg/mL) was used to achieve 50% BPR [35]. A previous report has documented 50% reduction in *L. monocytogenes* biofilm when pre-treated with DNase at concentration of 100 µg/mL and 75% reduction in case of post-treatment on 72 h old biofilm with contact duration of 24 h [22]. The basis for higher BPR achieved in this study further alludes that, in *P. aeruginosa* PAO1, e-DNA has a very crucial role to play in the process of biofilm formation [11,12].

Another valuable outcome of the current study is the synergistic effect of Mg^{2+} ions on efficacy of DNase I against *P. aeruginosa* PAO1 biofilms. Introduction of Mg^{2+} reduced the effective concentration of DNase I by 85% (reduction from 10 to 1.5 µg/mL) to achieve 80% and 75% BPR on polystyrene and polypropylene, respectively. We could not come across any study wherein introduction of Mg^{2+} has led to such a drastic increase in efficacy of DNase I. However, studies are available wherein introduction of Mg^{2+} has been reported to restore the antibiofilm effect of DNase I against *P. aeruginosa* biofilms [36]. Divalent ion, Mg^{2+} is a cofactor of DNase I and their addition seems to have improved the efficacy of the enzyme. These ions, however, have also been reported to reduce the efficacy of antibiotics [37,38]. Therefore, further studies are required for coming up with a strategy encompassing antimicrobials, DNase and Mg^{2+} ions for effective control of biofilms.

DNase I pre-treatment, post treatment and dual treatment (combination of pre-treatment and post treatment) on individual and mixed species biofilm revealed very interesting results. There was an insignificant difference in the effect of treatments on individual biofilms. The susceptibility of biofilms to DNase I was organism specific. These findings indicate that the biofilms vary with respect to their dependence on e-DNA for biofilm formation. To the best of the authors' knowledge, there is a lack of reports wherein multiple pathogens have been compared in reference to antibiofilm effect of DNase I treatments (pre, post and dual); therefore, the findings could not be corroborated.

The optimized DNase I treatments were tested against mixed species biofilms, which is a more accurate simulation of biofilms in real-life scenarios. The biofilms formed by test organisms individually were greater than that formed when the respective organism was dominant (2× inoculum) in mixed species. This observation can be attributed to the competitive and/or antagonistic interaction in the mixed species biofilm [39]. On the contrary, synergism amongst biofilm inhabitants has also been reported [30,40]. Overall, in the mixed species biofilm, DNase I was not as effective as against individual biofilms. The post treatments of mixed species biofilms grown for 24 h led to BPR in the range of 36–76%, which further declined to 13–53% with the ageing of biofilm for 48 h, except *Kelbsiella* 2× biofilms. Overall, mixed species biofilm is explored to lesser extent than individual biofilms and therefore we could not come across any study, wherein biofilms of more than two organisms were developed and treated with DNase I. The findings of the current investigation are slightly better than published studies on dual species biofilm of *Candida albicans* and *S. epidermidis* and *C. albicans* and *Streptococcus gordonii*, which have reported BPR of 35% and 25% [41,42]. Others have reported 45% and 80% reduction in viable cell count in dual species 48 h old biofilm formed by *L. monocytogenes*—*E. coli* and *L. monocytogenes*—*Pseudomonas fluorescens*, respectively, when post-treated with DNase (400 µg/mL, contact duration of 30 min) [43].

The prospect of using DNase I treatment as a part of clean-in-place regimes in the food industry are bolstered by the fact that it is heat sensitive and would be deactivated during heat treatments deployed in the food industry [44,45]. Moreover, if ingested along with the food items, acidic pH prevalent in stomach will degrade the enzyme [45]. DNase I based human therapeutics agents are also being developed for cystic fibrosis and rapid wound healing [46–48]. DNase I coating on polymethylmethacrylate biomaterial has been suggested for effective antibiotic delivery [49]. These reports indicate that time has ripened for the development of DNase based antibiofilm formulations for the food industry.

5. Conclusions

In general, the findings of the current study indicate that post-treatment with DNase I was superior to pre-treatment and dual treatment even when applied to solo or mixed biofilms. In addition, DNase is effective to remove biofilms on various substrates used in the food industry like polypropylene and polystyrene. DNase itself is not an antimicrobial but can effectively sensitize the biofilm structure for antimicrobial. DNase can be considered for clean-in-place regimes in food industries in view of its efficacy in reducing biofilm formation or removing pre-existing biofilms. However, further research is required to understand the effect of DNase especially on mixed species biofilm in nature where conditions are not conducive for DNase.

Acknowledgments: The authors acknowledge the extramural funding provided by the Department of Biotechnology (DBT, New Delhi) under the "Bio-CARe" scheme. The authors are also thankful to Chotubhai Lallubhai Patel, Chairman, CharutarVidya Mandal, Vidya Nagar, Anand and Rajeev Kumar Jain, Principal, AD Patel Institute of Technology, New VallabhVidya Nagar for providing necessary infrastructure and dedicated lab space for research. We also express heartfelt gratitude to Shivmurti Srivastav, Head, Food Processing Technology, ADIT for constant support and encouragement.

Author Contributions: A.P.S. conceived and designed the experiments; K.S. performed the experiments; and the contribution of A.P.S. and K.S. for analyzing and writing the paper was 70:30, respectively.

Conflicts of Interest: The authors declare no conflict of interest.

References

1. Lebeaux, D.; Chauhan, A.; Rendueles, O.; Beloin, C. From in vitro to in vivo Models of Bacterial Biofilm-Related Infections. *Pathogens* **2013**, *2*, 288–356. [CrossRef] [PubMed]
2. Singh, A.P.; Singh, J. Antibiofilm strategies. In *Biofilms in Bioengineering*, 1st ed.; Simões, M., Mergulhão, F., Eds.; Nova Biomedical; Nova Science Publishers: Porto, Portugal, 2013; Volume 1, p. 363. ISBN 978-1-62948-161-6.
3. Lasarre, B.; Federle, M.J. Exploiting Quorum Sensing to Confuse Bacterial. *Pathogens* **2013**, *77*, 73–111. [CrossRef] [PubMed]
4. Chen, X.; Schauder, S.; Potier, N.; Van Dorsselaer, A.; Pelczer, I.; Bassler, B.L.; Hughson, F.M. Structural identification of a bacterial quorum-sensing signal containing boron. *Nature* **2002**, *415*, 545–549. [CrossRef] [PubMed]
5. Chang, H.; Zhou, J.; Zhu, X.; Yu, S.; Chen, L.; Jin, H.; Cai, Z. Strain identification and quorum sensing inhibition characterization of marine-derived *Rhizobium* NAO1. *R. Soc. Open Sci.* **2017**, *77*, 73–111. [CrossRef]
6. Pagedar, A.; Singh, J.; Batish, V.K. Efflux mediated adaptive and cross resistance to ciprofloxacin and benzalkonium chloride in *Pseudomonas aeruginosa* of dairy origin. *J. Basic Microbiol.* **2011**, *51*, 289–295. [CrossRef] [PubMed]
7. Pagedar, A.; Singh, J. Influence of physiological cell stages on biofilm formation by *Bacillus cereus* of dairy origin. *Int. Dairy J.* **2012**, *23*, 30–35. [CrossRef]
8. Flemming, H.C.; Wingender, J. Extracellular Polymeric Substances (EPS): Structural, Ecological and Technical aspects. In *Encyclopedia of Environmental Microbiology*; Bitton, G., Ed.; John Wiley & Sons: New York, NY, USA, 2002; pp. 1223–1231.
9. Montanaro, L.; Poggi, A.; Visai, L.; Ravaioli, S.; Campoccia, D.; Speziale, P.; Arciola, C.R. Extracellular DNA in biofilms. *Int. J. Artif. Organ.* **2011**, *34*, 824–831. [CrossRef] [PubMed]
10. Sahu, P.K.; Iyer, P.S.; Oak, A.M.; Pardesi, K.R.; Chopade, B. Characterization of eDNA from the Clinical Strain *Acinetobacter baumannii* AIIMS 7 and Its Role in Biofilm Formation. *Sci. World J.* **2012**, 1–10. [CrossRef] [PubMed]
11. Whitchurch, C.B.; Tolker-Nielsen, T.; Ragas, P.C.; Mattick, J.S. Extracellular DNA required for bacterial biofilm formation. *Science* **2002**, *295*, 1487. [CrossRef] [PubMed]
12. Fleming, D.; Rumbaugh, K. Approaches to Dispersing Medical Biofilms. *Microorganisms* **2017**, *5*, 15. [CrossRef] [PubMed]
13. Beenken, K.E.; Spencer, H.; Griffin, L.M.; Smelter, S.M.; Camilli, A. Impact of extracellular nuclease production on the biofilm phenotype of *Staphylococcus aureus* under in vitro and in vivo conditions. *Infect. Immun.* **2012**, *80*, 1634–1638. [CrossRef] [PubMed]
14. Gnanadhas, D.P.; Elango, M.; Datey, A. Chronic lung infection by *Pseudomonas aeruginosa* biofilm is cured by L-Methionine in combination with antibiotic therapy. *Sci. Rep.* **2015**, *5*, 16043. [CrossRef] [PubMed]
15. Izano, E.; Amarante, M.; Kher, W.B.; Kaplan, J.B. Differential roles of poly-N-acetylglucosamine surface polysaccharide and extracellular DNA in *Staphylococcus aureus* and *Staphylococcus epidermidis* biofilms. *Appl. Environ. Microbiol.* **2008**, *74*, 470–476. [CrossRef] [PubMed]
16. Kaplan, J.B.; Lovetri, K.; Cardona, S.T.; Madhyastha, S.; Sadovskaya, I.; Jabbouri, S.; Izano, E.A. Recombinant human DNase I decreases biofilm and increases antimicrobial susceptibility in staphylococci. *J. Antibiot.* **2012**, *65*, 73–77. [CrossRef] [PubMed]

17. Klare, W.; Das, T.; Ibugo, A.; Buckle, E.; Manefield, M.; Manos, J. Glutathione-disrupted biofilms of clinical *Pseudomonas aeruginosa* Strains Exhibit an Enhanced Antibiotic Effect and a Novel Biofilm. *Antimicrob. Agents Chemother.* **2016**, *60*, 4539–4551. [CrossRef] [PubMed]

18. Brown, H.L.; Reuter, M.; Hanman, K.; Betts, R.P.; Van Vliet, A.H.M. Prevention of biofilm formation and removal of existing biofilms by extracellular DNase of *Campylobacter jejuni*. *PLoS ONE* **2015**, *10*, e0121680. [CrossRef] [PubMed]

19. Brown, H.L.; Hanman, K.; Reuter, M.; Betts, R.P.; van Vliet, A.H.M. *Campylobacter jejuni* biofilms contain extracellular DNA and are sensitive to DNase I treatment. *Front. Microbiol.* **2015**, *6*, 1–11. [CrossRef] [PubMed]

20. Fuxman Bass, J.I.; Russo, D.M.; Gabelloni, M.L.; Geffner, J.R.; Giordano, M.; Catalano, M.; Zorreguieta, A.; Trevani, A.S. Extracellular DNA: A Major Proinflammatory Component of *Pseudomonas aeruginosa* Biofilms. *J. Immunol.* **2010**, *184*, 6386–6395. [CrossRef] [PubMed]

21. Tetz, G.V.; Artemenko, N.K.; Tetz, V.V. Effect of DNase and antibiotics on biofilm characteristics. *Antimicrob. Agents Chemother.* **2009**, *53*, 1204–1209. [CrossRef] [PubMed]

22. Nguyen, U.T.; Burrows, L.L. DNase I and proteinase K impair *Listeria monocytogenes* biofilm formation and induce dispersal of pre-existing biofilms. *Int. J. Food Microbiol.* **2014**, *187*, 26–32. [CrossRef] [PubMed]

23. Fredheim, E.G.A.; Klingenberg, C.; Rohde, H.; Frankenberger, S.; Gaustad, P.; Flaegstad, T.; Sollid, J.E. Biofilm formation by *Staphylococcus shaemolyticus*. *J. Clin. Microbiol.* **2009**, *47*, 1172–1180. [CrossRef] [PubMed]

24. Cavaliere, R.; Ball, J.L.; Turnbull, L.; Whitchurch, C.B. The biofilm matrix destabilizers, EDTA and DNaseI, enhance the susceptibility of nontypeable *Hemophilus influenzae* biofilms to treatment with ampicillin and ciprofloxacin. *Microbiol. Open* **2014**, *3*, 557–567. [CrossRef] [PubMed]

25. Kawarai, T.; Narisawa, N.; Suzuki, Y.; Nagasawa, R.; Senpuku, H. *Streptococcus mutans* biofilm formation is dependent on extracellular DNA in primary low pH conditions. *J. Oral Biosci.* **2016**, *58*, 55–61. [CrossRef]

26. Arslan, S.; Eyi, A.; Özdemir, F. Spoilage potentials and antimicrobial resistance of *Pseudomonas* spp. isolated from cheeses. *J. Dairy Sci.* **2011**, *94*, 5851–5856. [CrossRef] [PubMed]

27. Guidelines for Drinking-water Quality. Available online: http://apps.who.int/iris/bitstream/10665/254637/1/9789241549950-eng.pdf?ua=1 (accessed on 16 March 2018).

28. O'Toole, G. Microtiter Dish Biofilm Formation Assay. *J. Vis. Exp.* **2011**, 10–11. [CrossRef] [PubMed]

29. XLstatistics Add-in Tool for Excel. Available online: http://www.deakin.edu.au/~rodneyc/XLStatistics/ (accessed on 16 March 2018).

30. Jahid, I.K.; Ha, S.D. The Paradox of Mixed-Species Biofilms in the Context of Food Safety. *Comp. Rev. Food Sci. Food Saf.* **2014**, *13*, 990–1011. [CrossRef]

31. Sharma, K.; Singh, J.; Singh, A.P. Combating biofilm mediated antimicrobial resistance using efflux pump inhibitor and Deoxyribonuclease. *Int. J. Mgmt. Appl. Sci.* **2017**, *5*, 16–19.

32. Tetz, V.V.; Tetz, G.V. Effect of extracellular DNA destruction by DNase I on characteristics of forming biofilms. *DNA Cell Biol.* **2010**, *29*, 399–405. [CrossRef] [PubMed]

33. Alhede, M.; Kragh, K.N.; Qvortrup, K.; Allesen-Holm, M.; van Gennip, M.; Christensen, L.D.; Jensen, P.Ø.; Nielsen, A.K.; Parsek, M.; Wozniak, D.; et al. Phenotypes of non-attached *Pseudomonas aeruginosa* aggregates resemble surface attached biofilm. *PLoS ONE* **2011**, *6*, e27943. [CrossRef] [PubMed]

34. Schlafer, S.; Meyer, R.L.; Dige, I.; Regina, V.R. Extracellular DNA Contributes to Dental Biofilm Stability. *Caries Res.* **2017**, *51*, 436–442. [CrossRef] [PubMed]

35. Grande, R.; Giulio, M.; Bessa, L.J.; Di Campli, E.; Baffoni, M.; Guarnieri, S.; Cellini, L. Extracellular DNA in *Helicobacter pylori* biofilm: A backstairs rumour. *J. Appl. Microbiol.* **2011**, *110*, 490–498. [CrossRef] [PubMed]

36. Mulcahy, H.; Charron-Mazenod, L.; Lewenza, S. Extracellular DNA chelates cations and induces antibiotic resistance in *Pseudomonas aeruginosa* biofilms. *PLoS Pathog.* **2008**, *4*, e1000213. [CrossRef] [PubMed]

37. Zimelis, V.M.; Jackson, G.G. Activity of Aminoglycoside Antibiotics against *Pseudomonas aeruginosa* specificity and site of calcium and magnesium antagonism. *J. Infect. Dis.* **1973**, *127*, 663–669. [CrossRef] [PubMed]

38. Beware Mixing Antibiotics with Magnesium. Available online: https://www.newsmax.com/Health/Health-Wire/magnesium-supplements-antibiotics-mixing/2015/05/27/id/646969/ (accessed on 16 March 2018).

39. Rendueles, O.; Ghigo, J.M. Multi-species biofilms: How to avoid unfriendly neighbors. *FEMS Microbiol. Rev.* **2012**, *36*, 972–989. [CrossRef] [PubMed]

40. Varposhti, M.; Entezari, F.; Feizabadi, M.M. Synergistic interactions in mixed-species biofilms of pathogenic bacteria from the respiratory tract. *Rev. Soc. Bras. Med. Trop.* **2014**, *47*, 649–652. [CrossRef] [PubMed]

41. Pammi, M.; Liang, R.; Hicks, J.; Mistretta, T.A.; Versalovic, J. Biofilm extracellular DNA enhances mixed species biofilms of *Staphylococcus epidermidis* and *Candida albicans. BMC Microbiol.* **2013**, *13*, 257. [CrossRef] [PubMed]

42. Jack, A.A.; Daniels, D.E.; Jepson, M.A.; Vickerman, M.M.; Lamont, R.J.; Jenkinson, H.F.; Nobbs, A.H. *Streptococcus gordonii* com CDE (competence) operon modulates biofilm formation with *Candida albicans. Microbiology* **2015**, *161*, 411–421. [CrossRef] [PubMed]

43. Rodríguez-López, P.; Carballo-Justo, A.; Draper, L.A.; Cabo, ML. Removal of *Listeria monocytogenes* dual-species biofilms using combined enzyme-benzalkonium chloride treatments. *Biofouling* **2017**, *33*, 45–58. [CrossRef] [PubMed]

44. Deoxyribonuclease I from Bovine Pancreas, Product Information. Available online: https://www.sigmaaldrich.com/content/dam/sigma-aldrich/docs/Sigma/Product_Information_Sheet/dneppis.pdf (accessed on 16 March 2018).

45. Thermo Fisher Scientific. Available online: https://www.thermofisher.com/in/en/home/references/ambion-tech-support/nuclease-enzymes/general-articles/dnase-i-demystified.html (accessed on 28 October 2017).

46. Torbic, H.; Hacobian, G. Evaluation of Inhaled Dornase Alfa Administration in Non-Cystic Fibrosis Patients at a Tertiary Academic Medical Center. *J. Pharm. Pract.* **2016**, *29*, 480–483. [CrossRef] [PubMed]

47. Maxwell, R.E.; Loomis, E.C. Fibrinolysin-Desoxyribonuclease for Enzymatic Debridement. Available online: https://www.google.co.in/patents/US3208908 (accessed on 28 October 2017).

48. Hanson, D.P. Composition for Enzymatic Debridement. Available online: https://www.google.com/patents/US20130156745 (accessed on 28 October 2017).

49. Swartjes, J.J.; Das, T.; Sharifi, S.; sharifi, S.; Subbiahdoss, G.; Sharma, P.K.; Krom, P.B.; Busscher, H.J.; Vander Mei, H.C. A functional DNase I coating to prevent adhesion of bacteria and the formation of biofilm. *Adv. Funct. Mater.* **2013**, *23*, 2843–2849. [CrossRef]

foods

Article

Effect of Food Residues in Biofilm Formation on Stainless Steel and Polystyrene Surfaces by *Salmonella enterica* Strains Isolated from Poultry Houses

Alba María Paz-Méndez, Alexandre Lamas *, Beatriz Vázquez, José Manuel Miranda, Alberto Cepeda and Carlos Manuel Franco

Laboratorio de Higiene, Inspección y Control de Alimentos, Dpto. de Química Analítica, Nutrición y Bromatología, Universidad de Santiago de Compostela, 27002 Lugo, Spain; albamaria.paz@rai.usc.es (A.M.P.-M.); beatriz.vazquez@usc.es (B.V.); josemanuel.miranda@usc.es (J.M.M.); alberto.cepeda@usc.es (A.C.); carlos.franco@usc.es (C.M.F.)
* Correspondence: alexandre.lamas@usc.es; Tel.: +34-982-822-407

Received: 19 September 2017; Accepted: 27 November 2017; Published: 29 November 2017

Abstract: *Salmonella* spp. is a major food-borne pathogen around the world. The ability of *Salmonella* to produce biofilm is one of the main obstacles in reducing the prevalence of these bacteria in the food chain. Most of *Salmonella* biofilm studies found in the literature used laboratory growth media. However, in the food chain, food residues are the principal source of nutrients of *Salmonella*. In this study, the biofilm formation, morphotype, and motility of 13 *Salmonella* strains belonging to three different subspecies and isolated from poultry houses was evaluated. To simulate food chain conditions, four different growth media (Tryptic Soy Broth at 1/20 dilution, milk at 1/20 dilution, tomato juice, and chicken meat juice), two different surfaces (stainless steel and polystyrene) and two temperatures (6 °C and 22 °C) were used to evaluate the biofilm formation. The morphotype, motility, and biofilm formation of *Salmonella* was temperature-dependent. Biofilm formation was significantly higher with 1/20 Tryptic Soy Broth in all the surfaces and temperatures tested, in comparison with the other growth media. The laboratory growth medium 1/20 Tryptic Soy Broth enhanced biofilm formation in *Salmonella*. This could explain the great differences in biofilm formation found between this growth medium and food residues. However, *Salmonella* strains were able to produce biofilm on the presence of food residues in all the conditions tested. Therefore, the *Salmonella* strain can use food residues to produce biofilm on common surfaces of the food chain. More studies combining more strains and food residues are necessary to fully understand the mechanism used by *Salmonella* to produce biofilm on the presence of these sources of nutrients.

Keywords: *Salmonella*; biofilm; morpothypes; stainless steel; food residues; tomato; poultry; milk

1. Introduction

Salmonella spp. are major food-borne pathogens around the world. The *Salmonella* genus is composed by two species, *S. bongori* and *S. enterica*. Also, the latter is also composed of six subspecies: *S. enterica* (I), *S. salamae* (II), *S. arizonae* (IIIa), *S. diarizonae* (IIIb), *S. houtenae* (IV), and *S. indica* (VI) [1]. In the year 2015, *S. enterica* was responsible of 94,625 confirmed cases of salmonellosis and 126 deaths in the European Union (EU). Although in the last decade the cases of human salmonellosis followed a negative trend, the last report of the European Food Safety Agency (EFSA) showed a slight increase in the number of infections [2]. These results reveal the importance of continuing developing new strategies to avoid the persistence of *Salmonella* strains through the food supply chain. For this purpose, it is of great importance to fully understand the survival mechanism of this pathogen in the different environments of the food chain. The sources of *Salmonella* in the food chain are mainly poultry products such as chicken meat and eggs [2]. However, in the last years, fresh products such as vegetables have been also responsible of salmonellosis outbreaks due to, among other things, the use of polluted irrigation water. The presence of *Salmonella* strains in fresh products is a major public health problem as preservatives are not commonly used in these products and they are normally consumed raw [3,4].

One of the most important persistence mechanisms of *Salmonella* is biofilm formation. A biofilm is defined as a community of microorganisms of the same or different species enclosed in a self-produced polymeric matrix adhered to different kinds of live or abiotic surfaces [5]. Biofilms cells are characterized by an increased resistance to environmental stresses (i.e., UV radiation, pH change, osmotic shock, and desiccation), antimicrobials, biocides, and the host immune system in comparison with planktonic cells. Extracellular polymeric substances (EPSs) are one of the factors responsible of this protective effect [6]. The main EPSs in *Salmonella* biofilms are cellulose and curli fimbriae, whose combined production is the responsible of the RDAR (red, dry, and rough) morphotype. Although the relation between *Salmonella* virulence and RDAR morphotype is still unclear, it is demonstrated that *Salmonella* strains showing RDAR morphotype have a great ability to produce biofilm on abiotic surfaces [5,7]. The production of cellulose and curli fimbriae is closely related to the *csgD* and *adrA* genes. In the first place, the transcription of *csgD* results in the synthesis of the biofilm master regulator CsgD that directly activates the curli fimbriae biosynthesis genes and positively regulates the production of AdrA that activate the synthesis of cellulose through the *bcsA* gene [5]. *S. enterica* biofilm formation has been studied in a wide range of strains from different sources and under multiple environmental conditions being biofilm formation strain-dependent [8,9]. In addition, temperatures, nutrients, or oxygen levels highly influenced the amount of biofilm formed in *Salmonella* strains and the morphotype produced [10,11]. Therefore, the transcription of biofilm-related genes in *S. enterica* is closely related to the environmental conditions [12,13]. However, most of the studies carried out until now used laboratory growth media for biofilm formation studies. The results obtained in that kind of studies are only approximate because lab media have a well-balanced nutritional composition and do not represent the complex composition of the food products found in the food chain. For example, a recent study observed that growth media supplemented with meat juices residues increased biofilm formation in *S.* Typhimurium. Therefore, meat juices residues may act as a surface conditioner to support initial attachment to abiotic surfaces [14].

In this context, the aim of this study is to evaluate how food residues can influence the biofilm-forming ability of *S. enterica* strains belonging to three different subspecies and isolated from poultry houses. A common growth laboratory medium was used as a reference media. Tomato juice (vegetable industry), chicken juice (meat industry), and milk (dairy industry) were used as representations of the different products that can be processed in the food industry. In addition, to represent the different conditions that the strains can find in the different steps of the food chain, two surfaces (polystyrene and stainless steel) were tested in biofilm assays. Also, the morphotype and motility of all the strains in all the temperatures tested were determined.

2. Materials and Methods

2.1. Bacterial Strains and Growth Media

A total of 13 *Salmonella* strains belonging to three different subspecies of *S. enterica* were used in this study (Table 1). *Salmonella* strains were isolated from samples recollected from poultry houses as previously described [15]. The Kauffman–Whyte typing scheme for the detection of somatic (O) and flagellar (H) antigens, with standard antisera (Bio-Rad Laboratories, Hercules, CA, USA) was used to serotype *Salmonella* strains. *Salmonella* stock cultures were maintained at −20 °C in cryovials (Deltalab, Barcelona, Spain). These strains were revitalized by transferring one bead into 10 mL of Trypic Soy Broth (TSB, Oxoid, UK) and incubating for 24 h at 37 °C (precultures). To obtain the working cultures, 20 µL of *Salmonella* strains precultures were transferred into 10 mL of TSB and incubated 24 h at 37 °C.

Four different growth media were used for biofilm assays. TSB at 1/20 (w/w) was used as a growth laboratory reference media. The nutrient balance of food residues found by *Salmonella* in the food chain is not as adequate as common laboratory growth media and therefore can have deficiency of some important components. In this sense, 1/20 TSB is a nutrient-limited medium that has demonstrated to be effective in promoting biofilm formation in *Salmonella* [8,16]. Due to this characteristic, 1/20 TSB was the growth medium chosen for comparative purposes. To represent possible food residues found in the food processing industry, tomato and chicken meat juice and UHT milk diluted 1/20 (w/w) were used in the biofilm assays. These assays were carried out at two different temperatures (6 °C and 22 °C) and two different surfaces (polystyrene and stainless steel).

2.2. Tomato and Chicken Meat Juice Preparation

Chicken meat juice (CMJ) was obtained as previously described by Birk et al. [17]. Briefly, chicken was obtained from local supermarkets and frozen for 2 days at −20 °C. Then, chicken was placed in a plastic bucket and thawed overnight. Chicken juice was collected in microtubes of 1.5 mL and centrifuged at 10,000× g for 10 min to eliminate large particles. The supernatant was filtered using 0.45 μm filter and stored at −20 °C until use. To obtain tomato juice (TJ), 50 g of tomato was mixed with 50 mL of distilled water in a bag and homogenized for 2 min. The liquid obtained was transferred to 1.5 mL microtubes and centrifuged at 10,000× g for ten minutes and the supernatant was filtered with 0.45 μm filters and stored at −20 °C until use.

2.3. Polystirene Biofilm Formation Assays

The determination of the biofilm formation in polystyrene with the different growth media was measured at 6 °C and 22 °C. Assays were carried out according Stepanovic et al. [16] with some modifications. Briefly, 96-well polystyrene microplates were filled with 200 μL of growth medium, and 20 μL of *Salmonella* culture containing 10^8 CFU/mL after 24 h of incubation was added to each well. Then, the microplates were incubated under the tested temperatures for 48 h. After incubation, the liquid of the plate was poured off and the wells were washed three times with 300 μL of distilled water. *Salmonella* cells attached to the microplate walls were fixed using 250 μL of absolute methanol for 15 min and then the plates were emptied and air-dried. The wells were stained with 250 μL for 5 min with 0.1% crystal violet solution. Crystal violet was rinsed off by placing the microplate under running water. The microplates were air-dried, and the dye bound to the adherent bacterial cells was resolubilized using 250 μL of 33% glacial acetic acid. The optical density (OD) was measured at 630 nm with a Plate Reader (das, Roma, Italy). The assays were performed in triplicate in three independent experiments.

2.4. Stainless Steel Biofilm Formation Assays

Stainless steel coupons (3.5 × 3.5 cm) were used to determine biofilm formation of *Salmonella* strains with the different media and temperature tested in this study. The method used was an adaption based on Stepanovic et al. [16]. Briefly, the stainless steel coupons were placed at the bottom of 125 mL bottles (Deltalab, Spain) filled with 10 mL of the appropriate medium and 100 μL of overnight *Salmonella* culture containing 10^8 CFU/mL. These bottles were incubated under the tested temperatures for 48 h. To remove non-adhered cells, the stainless steel coupons were washed with 10 mL of running distilled water, using a 10 mL micropipette. *Salmonella* attached to the stainless steel were fixed by immersing the coupons in absolute methanol for 15 min. After that, the coupons were air-dried and immersed in a 0.1% crystal violet solution for 5 min. The excess crystal violet was rinsed off by placing the stainless steel coupons under running water and air-drying. Finally, the coupons were placed in petri dishes containing 10 mL of 33% acetic acid to resolubilize the crystal violet. Finally, 200 μL of these solutions was poured in a 96-well microplate and the OD was measured at 630 nm with a plate reader (das, Roma, Italy).

2.5. Determination of Morphotype

The morphotype of the strains was determined at 6 °C and 22 °C as previously described by RömLing et al. [18] with some modifications. Briefly, TSB overnight *Salmonella* cultures were spread-plated onto Luria-Bertani (LB) plates without salt and supplemented with 40 mg/L of Congo red and 20 mg/L of Coomassie brilliant blue. The plates were incubated for 96 h and the morphotypes were determined in each strain at each temperature. The morphotypes in Congo red agar were classified as RDAR (red, dry, and rough; produce curli fimbriae and cellulose), SAW (soft and white), and SACW (soft and completely white; produce neither curli fimbriae nor cellulose and colonies were totally white).

2.6. Motility Assays

The motility of each strain in the different atmospheres was tested using a semisolid motility test medium according Karatzas et al. [19] with some modifications. The medium was composed by 10 g/L tryptone (Cultimed, Panreac, Barcelona, Spain), 5 g/L NaCl (Panreac, Barcelona, Spain), 4 g/L agar (Liofilchem, Roseto degli Abruzzi, Italy), 3 g/L beef extract (Oxoid Ltd., Thermo Scientific, Hampshire, UK), and 0.05 g/L of 2,3,5 triphenyltetrazolium chloride (Sigma-Aldrich, Taufkirchen, Germany), and was sterilized (15 min at 121 °C). Overnight cultures in TSB were transferred to the motility agar by stabbing. The plates were incubated at 22 °C and 6 °C for 72 h. *Salmonella* metabolism produces a red color when swimming away in the motility agar due to the reduction of 2,3,5 triphenyltetrazolium chloride to formazan. Finally, the ratio between the inoculum site and the edge of the red circle was measured as an indication of the motility.

2.7. Statistical Anaylisis

Statistical analyses were carried out with SPSS software for Windows (SPSS Inc., Chicago, IL, USA). Analysis of variance (ANOVA) was used to study the influence of growth media and temperature of incubation in the biofilm formation ability of *Salmonella* strains.

3. Results

A total of 13 strains isolated from poultry houses were used in this study (Table 1). The morphotype of *Salmonella* strains was evaluated under two different temperatures (6 °C and 22 °C). The results showed that *Salmonella* strains produced different morphotypes at different temperatures of incubation. All the strains produced the RDAR morphotype at 22 °C with the exception of *S. enterica* subsp. *arizonae* strains Lhica 2 and Lhica 6, which produced the SAW morphotype at 22 °C. However, at 6 °C, all the *Salmonella* strains tested in this study produced a morphotype characterized to be totally white, and therefore this morphotype was called by the authors as soft and completely white (SACW) (Figure 1).

Figure 1. Morphotype Soft and Completely White (SCAW) produced by *Salmonella* strains at 6 °C.

Table 1. *Salmonella enterica* strains selected for this study and the morphotype produced under different incubation temperatures tested in this study (6 °C and 22 °C).

	Morphotype	
Serotype/subspecies	6 °C	22 °C
S. Typhimurium Lhica T1	SACW	RADR
S. Typhimurium Lhica T4	SACW	RADR
S. Typhimurium Lhica T5	SACW	RADR
S. Typhimurium Lhica T6	SACW	RADR
S. Enteritidis Lhica ET1	SACW	RADR
S. Bardo Lhica B2	SACW	RADR
S. Newport Lhica N5	SACW	RADR
S. Infantis Lhica I4	SACW	RADR
S. Infantis Lhica I5	SACW	RADR
S. enterica subsp. *arizonae* serovar 48:z4,z23,z32:-Lhica AZ2	SACW	SAW
S. enterica subsp. *arizonae* serovar 48:z4,z23:-Lhica AZ6	SACW	SAW
S. enterica subsp. *salamae* serovar 4,12:b:-Lhica SA3	SACW	RADR
S. enterica subsp. *salamae* serovar 6,8:g,m,t:-Lhica SA2	SACW	RADR

RDAR, red, dry, and rough; SAW, smooth and white; SACW, soft and completely white.

The motility of *Salmonella* strains is closely correlated with their ability to produce biofilm. In this context, the motility of the strains used in this study was evaluated at 6 °C and 22 °C (Table 2). The mean motility of the strains was significantly higher ($p < 0.05$) at 22 °C than at 6 °C. There were also significant differences between the strains. At 22 °C, the motilities of *S.* Typhimurium Lhica T5 (25.10 ± 2.10 mm) and *S.* Enteritidis (25.40 ± 3.20 mm) were significantly higher than the motilities of the other strains. *S.* Infantis I5 presented lower motility at 22 °C (12.00 ± 1.00 mm). However, strains *S.* Infantis Lhica I4 (6.00 ± 0.50 mm) and *S.* Typhimurium T4 (6.00 ± 2.00 mm) presented higher motility at 6 °C. It is remarkable that no motility was detected in *S.* Typhimurium T1 and *S.* Newport N5 at 6 °C.

Table 2. Main motility (mm) of *Salmonella* strains tested in this study at 22 °C and 6 °C. Asterisks indicate significant differences ($p < 0.05$) in each strain between the two temperatures.

Strain	Motility (mm) at 22 °C	Motility (mm) at 6 °C
S. Typhimurium Lhica T1	15.50 ± 0.50	0.00 ± 0.00 *
S. Typhimurium Lhica T4	21.00 ± 1.00	6.00 ± 1.00 *
S. Typhimurium Lhica T5	25.10 ± 2.10	0.83 ± 0.29 *
S. Typhimurium Lhica T6	19.66 ± 1.53	2.17 ± 0.29 *
S. Enteritidis Lhica ET1	25.40 ± 3.20	2.00 ± 1.00 *
S. Bardo Lhica B2	19.83 ± 1.26	1.33 ± 0.58 *
S. Newport Lhica N5	19.50 ± 1.50	0.00 ± 0.00 *
S. Infantis Lhica I4	15.50 ± 0.50	6.00 ± 0.50 *
S. Infantis Lhica I5	12.00 ± 1.00	1.23 ± 0.25 *
S. enterica subsp. *arizonae* serovar 48:z4,z23,z32:-Lhica AZ2	20.07 ± 0.51	0.73 ± 0.40 *
S. enterica subsp. *arizonae* serovar 48:z4,z23:-Lhica AZ6	19.90 ± 0.79	1.00 ± 0.50 *
S. enterica subsp. *salamae* serovar 4,12:b:-Lhica SA3	15.33 ± 0.76	0.83 ± 0.57 *
S. enterica subsp. *salamae* serovar 6,8:g,m,t:-Lhica SA2	15.17 ± 1.04	1.83 ± 0.29 *
Average	18.62 ± 3.97	1.77 ± 1.96 *

Biofilm formation by *Salmonella* strains used in this study was evaluated under two different temperatures (6 °C and 22 °C), two different surfaces (polystyrene and stainless steel), and four different growth media (1/20 TSB, 1/20 Milk, Tomato juice, and Chicken meat juice). Table 3 shows the mean OD_{630} values obtained for each strain in the biofilm assays in polystyrene at 6 °C and 22 °C. All the strains produced biofilm with 1/20 TSB, 1/20 milk, and CMJ in both temperatures. However, with TJ not all the strains produced biofilm. The cutoff value of 0.070 was established to consider biofilm formation by the strains tested. This cutoff value was calculated according the

OD_{630} values obtained for the negative control wells (growth medium without strain) in polystyrene plates. Therefore, the strains that showed OD_{630} values lower than 0.070 in polystyrene assays were considered as not biofilm formed in those conditions. It is interesting that at 6 °C only one strain (*S.* Typhimurium Lhica T6) did not produce biofilm on TJ, and at 22 °C a total of five strains did not produce biofilm on TJ. In both temperatures, the mean OD_{630} average was significantly higher ($p < 0.05$) with 1/20 TSB than with the other growth media.

While there were no significant differences ($p = 0.110$) in the OD between 1/20 Milk, TJ, and CMJ at 6 °C, the growth medium 1/20 Milk showed significant higher ($p < 0.05$) OD values than the others two growth media at 22 °C. It is remarkable that both *S. enterica* subsp. *salamae* strains showed the higher OD_{630} values both in 1/20 TSB and 1/20 Milk media. It is especially interesting in the case of *S. enterica* subsp. *salamae* Lhica SA2 at 6 °C, which showed an OD_{630} value three times higher than *S.* Bardo Lhica B2, the non-*salamae* strain that showed the higher OD_{630} value.

The OD_{630} values obtained in stainless steel assays are not directly compared with those of polystyrene assays. Although the scientific principle is the same for both methods, they have slight differences in the quantities of reagents used. As in the case of polystyrene, the mean OD_{630} values were significantly higher ($p < 0.05$) with 1/20 TSB in both temperatures (6 °C and 22 °C). Between the other three growth media used, there were no significant differences (Table 4) at 6 °C and 22 °C, with the exception of TJ at 22 °C, where OD_{630} values were significantly lower ($p < 0.05$) than CMJ values.

While in polystyrene assays, *S. enterica* subsp. *salamae* strains presented the higher OD_{630} values in 1/20 TSB, in stainless steel assays the *S.* Enteritidis Lhica ET1 showed the higher OD_{630} values. The OD_{630} cutoff value established for stainless steel assays was 0.050. Therefore, all the strains were able to produce biofilm on stainless steel assays. All the strains presented higher OD_{630} values with 1/20 TSB, with the exception of *S.* Newport Lhica N5 at 6 °C, *S.* Typhimurium Lhica T6 at 22 °C, and *S. enterica* subsp. *arizonae* Lhica AZ6 at 6 °C and 22 °C, which presented higher values with CMJ. The incubation temperature also influenced the biofilm formation ability of the *Salmonella* strains tested in this study. In polystyrene and also in stainless steel, the mean OD_{630} values obtained with 1/20 TSB, 1/20 Milk, and CMJ were significantly higher at 22 °C than at 6 °C (Tables 3 and 4). However, biofilm formation in TJ was not influenced by temperature in polystyrene and stainless steel.

Table 3. Biofilm formation on polystyrene plates with the four different growth media used at 6 °C and 22 °C by *Salmonella* strains tested in this study.

| Strains | Media | | | | | | | |
| | 1/20 TSB | | 1/20 Milk | | TJ | | CMJ | |
St.	6 °C	22 °C	6 °C	22 °C	6 °C	22 °C	6 °C	22 °C
T1	0.112 ± 0.007 a	0.366 ± 0.089 a,*	0.092 ± 0.010 a,b	0.262 ± 0.077 b,*	0.079 ± 0.002 b	0.061 ± 0.004 d	0.083 ± 0.003 b	0.100 ± 0.015 c
T4	0.149 ± 0.012 a	0.395 ± 0.159 a,*	0.084 ± 0.011 b	0.139 ± 0.036 b,*	0.083 ± 0.007 b	0.059 ± 0.006 d,*	0.080 ± 0.004 b	0.088 ± 0.016 c
T5	0.146 ± 0.045 a	0.169 ± 0.057 a	0.099 ± 0.026 b	0.160 ± 0.089 a	0.078 ± 0.009 c	0.080 ± 0.005 c	0.077 ± 0.005 c	0.111 ± 0.013 b,*
T6	0.152 ± 0.012 a	0.219 ± 0.038 a,*	0.090 ± 0.009 b	0.144 ± 0.052 b,*	0.057 ± 0.005 d	0.071 ± 0.010 d	0.076 ± 0.004 c	0.113 ± 0.018 c,*
ET1	0.112 ± 0.007 a	0.461 ± 0.069 a,*	0.101 ± 0.014 a,b	0.199 ± 0.032 b,*	0.082 ± 0.008 b,c	0.088 ± 0.010 c	0.079 ± 0.010 c	0.091 ± 0.015 c,*
B2	0.232 ± 0.037 a	0.452 ± 0.042 a,*	0.108 ± 0.026 b	0.160 ± 0.033 b,*	0.080 ± 0.010 c	0.116 ± 0.012 c,*	0.085 ± 0.003 c	0.156 ± 0.022 b
N5	0.201 ± 0.022 a	0.424 ± 0.049 a,*	0.105 ± 0.017 b	0.191 ± 0.035 b,*	0.110 ± 0.013 b	0.081 ± 0.002 c	0.097 ± 0.025 b	0.105 ± 0.034 c
I4	0.076 ± 0.008 a	0.131 ± 0.025 a,*	0.104 ± 0.039 a	0.117 ± 0.061 a	0.100 ± 0.021 a	0.118 ± 0.032 a	0.111 ± 0.039 a	0.116 ± 0.025 a
I5	0.158 ± 0.012 a	0.404 ± 0.135 a,*	0.104 ± 0.021 b	0.146 ± 0.021 b	0.082 ± 0.003 b	0.125 ± 0.022 b,*	0.079 ± 0.004 c	0.085 ± 0.022 c
AZ3	0.115 ± 0.013 a	0.142 ± 0.063 a	0.080 ± 0.010 b	0.089 ± 0.020 b	0.083 ± 0.005 b	0.060 ± 0.009 c,*	0.079 ± 0.006 b	0.089 ± 0.019 b
AZ6	0.131 ± 0.015 a	0.098 ± 0.011 a,*	0.087 ± 0.020 b	0.100 ± 0.018 a	0.090 ± 0.010 b	0.067 ± 0.010 b,*	0.082 ± 0.007 c	0.097 ± 0.012 a
SA2	0.642 ± 0.098 a	0.563 ± 0.034 a	0.130 ± 0.040 b	0.300 ± 0.060 b,*	0.079 ± 0.010 c	0.081 ± 0.011 d	0.080 ± 0.020 d	0.121 ± 0.022 c,*
SA3	0.433 ± 0.019 a	0.566 ± 0.061 a,*	0.124 ± 0.032 b	0.314 ± 0.039 b,*	0.081 ± 0.001 c	0.066 ± 0.009 d,*	0.077 ± 0.012 c	0.096 ± 0.009 c,*
X̄	0.210 ± 0.161 a	0.341 ± 0.165 a,*	0.101 ± 0.020 b	0.179 ± 0.070 b,*	0.084 ± 0.015 b	0.083 ± 0.023 c	0.081 ± 0.014 b	0.105 ± 0.022 c,*

Different letters in the same row and the same temperature indicate significant differences ($p < 0.05$). Asterisks indicate significant differences ($p < 0.05$) between the two temperatures for the same growth medium. Strains code corresponds to those in Table 1. St.: strains; X̄: Average. Tryptic Soy Broth 1/20 at 1/20 (w/w) (1/20 TSB); Milk at 1/20 (w/w) (1/20 Milk); Tomato Juice (TJ); Chicken meat juice (CMJ).

Table 4. Biofilm formation on stainless steel coupons with the four different growth media used at 6°C and 22°C by *Salmonella* strains tested in this study.

| Strains | Media | | | | | | | |
| | 1/20 TSB | | 1/20 Milk | | TJ | | CMJ | |
St.	6 °C	22 °C	6 °C	22 °C	6 °C	22 °C	6 °C	22 °C
T1	0.161 ± 0.011 a	0.212 ± 0.010 a,*	0.098 ± 0.010 c	0.118 ± 0.008 b	0.090 ± 0.015 c	0.080 ± 0.010 c	0.123 ± 0.011 b	0.127 ± 0.012 b
T4	0.182 ± 0.009 a	0.228 ± 0.010 c	0.080 ± 0.008 c	0.107 ± 0.012 b	0.083 ± 0.006 c	0.111 ± 0.013 b	0.131 ± 0.005 b	0.112 ± 0.011 b
T5	0.134 ± 0.010 a	0.172 ± 0.010 a,*	0.076 ± 0.007 b	0.106 ± 0.012 b,*	0.073 ± 0.015 b	0.079 ± 0.012 c	0.071 ± 0.028 b	0.101 ± 0.013 b,*
T6	0.124 ± 0.007 a	0.124 ± 0.007 a,b	0.089 ± 0.010 b	0.111 ± 0.009 b,*	0.060 ± 0.004 c	0.056 ± 0.011 c	0.120 ± 0.014 a	0.136 ± 0.020 a
ET1	0.199 ± 0.010 a	0.247 ± 0.008 a,*	0.077 ± 0.010 b	0.109 ± 0.008 b,*	0.081 ± 0.009 c	0.069 ± 0.009 b	0.083 ± 0.019 b	0.107 ± 0.009 b
B2	0.141 ± 0.012 a	0.182 ± 0.008 a,*	0.100 ± 0.013 b	0.121 ± 0.009 b,*	0.059 ± 0.010 c	0.079 ± 0.010 c	0.104 ± 0.009 b	0.134 ± 0.011 b,*
N5	0.122 ± 0.011 a	0.162 ± 0.009 a,*	0.090 ± 0.009 b	0.122 ± 0.008 b,*	0.087 ± 0.010 b	0.089 ± 0.012 c	0.139 ± 0.021 a	0.150 ± 0.009 a
I4	0.080 ± 0.012 a	0.130 ± 0.011 a,*	0.082 ± 0.010 a	0.110 ± 0.008 a,*	0.060 ± 0.008 b	0.067 ± 0.005 b	0.085 ± 0.010 b	0.119 ± 0.010 a,*
I5	0.178 ± 0.002 a	0.218 ± 0.012 a,*	0.080 ± 0.010 c	0.104 ± 0.008 c,*	0.078 ± 0.012 c	0.055 ± 0.015 d,*	0.119 ± 0.012 b	0.139 ± 0.009 b,*
AZ3	0.099 ± 0.011 a	0.121 ± 0.009 a,*	0.101 ± 0.021 a	0.111 ± 0.010 a	0.092 ± 0.022 a	0.069 ± 0.012 b	0.099 ± 0.010 a	0.119 ± 0.005 a,*
AZ6	0.155 ± 0.006 a	0.131 ± 0.010 a,*	0.099 ± 0.018 a	0.101 ± 0.008 b	0.100 ± 0.020 a	0.070 ± 0.005 c,*	0.102 ± 0.030 a	0.139 ± 0.026 a,*
SA2	0.155 ± 0.015 a	0.198 ± 0.023 a,*	0.103 ± 0.014 b	0.110 ± 0.010 b,c	0.110 ± 0.020 b	0.098 ± 0.011 c	0.108 ± 0.020 b	0.135 ± 0.011 b,*
SA3	0.159 ± 0.012 a	0.188 ± 0.010 a,*	0.100 ± 0.006 b	0.115 ± 0.009 b	0.099 ± 0.010 b	0.120 ± 0.010 b	0.104 ± 0.016 b	0.129 ± 0.013 b
X̄	0.141 ± 0.036 a	0.178 ± 0.042 a,*	0.090 ± 0.010 b	0.111 ± 0.012 b,c,*	0.088 ± 0.016 b	0.084 ± 0.020 c	0.107 ± 0.035 b	0.127 ± 0.014 b,*

Different letters in the same row and the same temperature indicate significant differences ($p < 0.05$). Asterisks indicate significant differences ($p < 0.05$) between the two temperatures for the same growth medium. Strains code corresponds to those in Table 1. St.: strains; X̄: Average.

4. Discussion

The morphotype produced by *Salmonella* strains is closely related with the ability to produce biofilm. In this sense, the *Salmonella* morphotype RDAR, characterized by the production of cellulose and curli fimbriae, is produced as a mechanism of resistance to environmental conditions [5]. Thus, it was observed that *Salmonella* turned off the genetic machinery related with the production of RDAR morphotype during in vivo infection and turned on this machinery when *Salmonella* was in the external environment again [20]. Most studies used temperatures of 28 °C or higher to evaluate the morphotypes produced by *Salmonella* strains. However, a study carried out by Lamas et al. [10] with *Salmonella* strains isolated from poultry observed that the morphotype produced by *Salmonella* strains varied with the different temperatures tested. While at 37 °C, most of the strains produced the SAW morphotype, at 20 °C most of the strains produced the RDAR morphotype. Nevertheless, it is also important to evaluate the effect of refrigeration temperatures in biofilm morphotype production. In this study, it was observed that, with the exception of *S. enterica* subsp. *arizonae* strains, all the strains produced the RDAR morphotype at 22 °C. By contrast, at 6 °C all the strains used in this study produced a morphotype not previously described in the literature to the best of our knowledge. This morphotype is characterized to present as totally white (Figure 1). Due to this characteristic, this morphotype was described by the authors as soft and completely white (SACW). A direct relationship between the morphotype produced by *Salmonella* strains at different temperatures and the biofilm formation in the different growth media cannot be established because morphotype was determined only in a specific medium (LB without NaCl and with Congo Red and Coomassie brilliant blue) for this determination. In addition to temperature, growth media can influence the morphotype produced by *Salmonella* strains. It is remarkable that the case of *S. enterica* subsp. *salamae* Lhica SA2 produced more biofilm on polystyrene at 6 °C than at 22 °C with 1/20 TSB. It is possible that the different food residues influence the production of cellulose and curli fimbria at refrigerated temperatures. On the other hand, the production of RDAR is not totally essential for the production of higher amounts of biofilm. For example, previous studies carried out by Seixas et al. [21] and Solomon et al. [8] found no differences in the amount of biofilm produced by RDAR and SAW morphotypes. It is possible that *Salmonella* strains at refrigerated temperatures activate other genetic mechanisms related with biofilm formation, as it could be the production of colanic acid or maybe the flagella that plays an essential role at these temperatures.

Motility mediated by flagella has an important role in *Salmonella* persistence and colonization. Also, it has been observed that motility contributes the internalization of *Salmonella* into host and plant cells [22,23]. The role of flagella in biofilm formation is not totally clarified, but flagella seems to be important for the initial attachment step to surfaces and not for biofilm maturation [5]. The results of this study showed that temperature highly influenced the biofilm formation. With the exception of TJ, biofilm formation was higher at 22 °C than at 6 °C in the growth media tested (Tables 3 and 4). In the same way, the motility was significantly higher at 22 °C (Table 2). Low temperatures slow down the growth of microorganisms and may cause modifications in their metabolism that reduced their ability to produce biofilm. In this sense, it is possible that low temperatures reduced the synthesis of flagella in *Salmonella* cells, resulting in lower motility and lower capacity to attach to surfaces.

Previous studies [9–11,24] used growth laboratory media to perform biofilm assays. In the food industry, food residues are the principal source of nutrients used by food-borne pathogens. In an attempt to reproduce real conditions, this study used *Salmonella* strains isolated from the poultry industry to compare a common growth medium with different food residues on two common surfaces found in the food industry, polystyrene and stainless steel. For example, in poultry farms, polystyrene is used in water suppliers, feeding stations, or in the containers where living broilers are transported to the slaughterhouses. In food packaging, polystyrene is commonly used in chicken and beef packaging or in fruit packaging [25]. The results of this study showed that in both polystyrene and stainless steel, the mean OD_{630} observed was significantly higher with the 1/20 TSB growth medium in comparison with the other growth media tested. Therefore, it is possible that the results of biofilm assays performed with 1/20 TSB enhanced the biofilm formation of *Salmonella* strains, and their capacity to produce biofilm on the presence of food residues is lower in comparison with 1/20 TSB. In contrast with these results, a research carried out by Li et al. [14] observed that *Salmonella* strains formed more biofilm with meat juice than with the common laboratory growth medium Mueller-Hinton (MH) at 37 °C in polystyrene and glass surfaces. These different results could be due to the conditions used in both studies. It has been observed that high nutrient concentration media, such as MH or TSB without dilution, combined with temperatures of 37 °C results in lower biofilm formation by *Salmonella* strains [10,26].

Related to this, the temperature is another factor that influences biofilm formation. With the exception of TJ, the mean OD_{630} was higher at 22 °C than at 6 °C in the growth media used in this study. Lamas et al. [10] observed that 20 °C was the temperature at which *Salmonella* strains produced more biofilm with the 1/20 TSB growth medium. The same results were observed in this study for the 1/20 Milk and CMJ growth media. However, there were no significant differences in TJ between 6 °C and 22 °C. In this sense, Koukkidis et al. [27] observed that salad juices highly influenced growth at refrigerated temperatures, motility, and biofilm formation in *Salmonella*. Therefore, it is possible that other factors influence biofilm formation with these growth media. It is remarkable that, with one exception, all the strains were able to produce biofilm on polystyrene and stainless steel surfaces with all the growth media tested in this study. Both polystyrene and stainless steel surfaces are commonly found in the different steps of the food chain. The combination of food residues and *Salmonella* cells in these surfaces could result in biofilm formation and therefore cross-contamination of food products in contact with these surfaces with *Salmonella*. These results are a major of concern in public health and reflect the importance of maintaining adequate hygiene and disinfection methods to avoid the presence of food-borne pathogens biofilms, both in domestic refrigerators and storage chillers.

Raw milk is characterized for its complex microbial community composed of a wide range of bacterial genera that are able to form biofilm [28,29]. Although the influence of milk in biofilm formation has been evaluated in food-borne pathogens, such as *Listeria monocytogenes* [29] or *Staphylococcus aureus* [30], and spoilage bacteria [31], to the best of our knowledge the influence of milk in *Salmonella* biofilm formation has still not been evaluated. In this study, all the strains tested formed biofilm on stainless steel and polystyrene at 6 °C and 22 °C with 1/20 milk. Also, in polystyrene at 22 °C, the mean OD_{630} average was significantly higher in 1/20 milk than in other food residue media. It is also remarkable that the OD_{630} values obtained for *S. enterica* subsp. *salamae* strains at 22 °C in polystyrene were three times higher for 1/20 TSB than for the growth media TJ and CMJ. Therefore, this study demonstrates that *Salmonella* strains are able to produce biofilm on the presence of milk in both temperatures and surfaces. These results highlight the importance of good hygiene and disinfection practices in dairy equipment such as bulk tanks. Quorum-sensing molecules can increase biofilm formation in *Salmonella*, and this cell mechanism could play an important role for bacterial communication in multispecies biofilms [32,33]. Due to the microbiota composition of raw milk, *Salmonella* cells can integrate in biofilms formed by other bacterial genera as a response to quorum-sensing molecules produced by other microorganisms and liberated to the milk.

The strains used in this study were isolated from poultry houses, and all these strains were able to produce biofilm on all the surfaces and temperatures tested in this study with the growth medium CMJ. This result indicates that meat juice is a nutrient source for *Salmonella* in the food processing environment, allowing their biofilm formation. Different to polystyrene, in the stainless steel surface the mean OD_{630} value was higher in CMJ assays than in TJ and 1/20 milk assays at 6 °C and 22 °C. Therefore, it is possible that some compounds of CMJ facilitate the *Salmonella* biofilm formation on this surface. In this sense, it has been proposed that chicken meat juice used in laboratory could present residual quorum-sensing molecules that enhance the biofilm formation in food-borne pathogens. Also, Li et al. [14] observed that aflagellated mutants of *Campylobacter* and *Salmonella* increased their biofilm-forming ability when surfaces were pre-coated with a meat juice layer. Thus, the particles of meat juice could promote the initial attachment of *Salmonella* cells to inert contact surfaces and allow biofilm formation.

Salmonella strains have been related with fruit, nut, or vegetable outbreaks [34–36]. In this sense, *Salmonella* contamination of vegetables can be originated by contaminated composed manure, soil, animals, or irrigation and wash water [37]. Therefore, the capacity of Salmonella strains from poultry houses to produce biofilms on the surface of vegetables is a major public health problem. Different studies have observed that *Salmonella* is able to produce biofilm on parsley, rocket leaves, lettuce cucumber, and tomatoes [38]. Koukkidis et al. [27] observed that salad leaf juices enhanced the motility and biofilm produced by *Salmonella*. In this study, tomato juice was used as one growth medium to evaluate the effect of this extract in the ability of *Salmonella* strains to produce biofilm on polystyrene and stainless steel. All the *Salmonella* strains tested in this study produced biofilm on stainless steel at 6 °C and 22 °C. However, in polystyrene, five strains did not produce biofilm at 22 °C and one strain did not produce biofilm at 6 °C. Therefore, TJ seems to favor biofilm formation in stainless steel more than in polystyrene. Also, it is interesting that there were no significant differences between 6 °C and 22 °C in biofilm formation. It is possible that some tomato compounds improve biofilm formation at low temperatures. In this sense, the previously mentioned study by Koukkidis et al. [27] also observed that salad leaf juices enhance *Salmonella* growth at refrigeration temperatures. In addition, food residues such as carrot can have a protective effect in *Salmonella* cells' adherence to stainless steel [39]. The results of this study showed that *Salmonella* strains isolated from poultry houses produce biofilm on the presence of tomato residues.

5. Conclusions

Salmonella is one of the principal food-borne pathogens around the world. Authorities of different countries have adopted control strategies to reduce the prevalence of *Salmonella* in the food chain. However, the ability of *Salmonella* to produce biofilm is one of the main factors that make difficult their eradication from the food chain. Most of the studies carried out until now evaluated *Salmonella* biofilm formation using common laboratory growth media. However, biofilm formation is highly dependent on environmental conditions, and it is important to use food residues in biofilm assays to obtain results as close as possible to the real conditions of the food chain. The results of this study clearly showed that *Salmonella* strains isolated from poultry houses can produce biofilm both at 22 °C and 6 °C in stainless steel and polystyrene. Although biofilm formation was observed with all the growth media used, biofilm formation was significantly higher with the common laboratory growth medium. In this study, the laboratory growth medium 1/20 TSB was used, which enhances biofilm formation in *Salmonella*. This fact explains the high differences found between laboratory growth media and food residues. However, the effect of food residues in biofilm formation should not be underestimated. Future studies are necessary to confirm the results obtained in this study and to evaluate the effect of food residues in transcriptome of *Salmonella* cells. These data will allow the discovery of metabolic pathways involved in the interaction between *Salmonella* cells and food residues. Finally, it is necessary to develop standardized methods in biofilm assays to make possible the direct comparison of results obtained from different laboratories.

Author Contributions: A.L. and C.M.F. conceived and designed the experiments; A.M.P.-M. performed the experiments; A.L. and J.M.M. analyzed the data; B.V. contributed reagents, materials, and analysis tools; A.L. and A.C. wrote the paper.

Conflicts of Interest: The authors declare no conflict of interest.

References

1.	Grimont, P.A.; Weill, F. *Antigenic Formulae of the Salmonella Serovars*, 9th ed.; WHO Collaborating Centre for Reference and Research on Salmonella; WHO: Geneva, Switzerland, 2007.
2.	European Food Safety Authority. European Centre for Disease Prevention and Control the European Union summary report on trends and sources of zoonoses, zoonotic agents and food-borne outbreaks in 2015. *EFSA J.* **2016**, *14*, e04634.

3. Markland, S.; Ingram, D.; Kniel, K.; Sharma, M. Water for Agriculture: The Convergence of Sustainability and Safety. *Microbiol. Spectr.* **2017**, *5*. [CrossRef] [PubMed]

4. Wadamori, Y.; Gooneratne, R.; Hussain, M.A. Outbreaks and factors influencing microbiological contamination of fresh produce. *J. Sci. Food Agric.* **2017**, *97*, 1396–1403. [CrossRef] [PubMed]

5. Steenackers, H.; Hermans, K.; Vanderleyden, J.; De Keersmaecker, S.C. *Salmonella* biofilms: An overview on occurrence, structure, regulation and eradication. *Food Res. Int.* **2012**, *45*, 502–531. [CrossRef]

6. Hobley, L.; Harkins, C.; MacPhee, C.E.; Stanley-Wall, N.R. Giving structure to the biofilm matrix: An overview of individual strategies and emerging common themes. *FEMS Microbiol. Rev.* **2015**, *39*, 649–669. [CrossRef] [PubMed]

7. Pontes, M.H.; Lee, E.J.; Choi, J.; Groisman, E.A. *Salmonella* promotes virulence by repressing cellulose production. *Proc. Natl. Acad. Sci. USA* **2015**, *112*, 5183–5188. [CrossRef] [PubMed]

8. Solomon, E.B.; Niemira, B.A.; Sapers, G.M.; Annous, B.A. Biofilm formation, cellulose production, and curli biosynthesis by *Salmonella* originating from produce, animal, and clinical sources. *J. Food Prot.* **2005**, *68*, 906–912. [CrossRef] [PubMed]

9. De Oliveira, D.C.V.; Fernandes, J.A.; Kaneno, R.; Silva, M.G.; Araújo, J.J.P.; Silva, N.C.C.; Rall, V.L.M. Ability of *Salmonella* spp. to produce biofilm is dependent on temperature and surface material. *Foodborne Pathog. Dis.* **2014**, *11*, 478–483. [CrossRef] [PubMed]

10. Lamas, A.; Fernandez-No, I.C.; Miranda, J.M.; Vázquez, B.; Cepeda, A.; Franco, C.M. Biofilm formation and morphotypes of *Salmonella enterica* subsp. arizonae differs from those of other *Salmonella enterica* subspecies in isolates from poultry houses. *J. Food Prot.* **2016**, *79*, 1127–1134. [CrossRef] [PubMed]

11. Lamas, A.; Miranda, J.M.; Vázquez, B.; Cepeda, A.; Franco, C.M. Biofilm formation, phenotypic production of cellulose and gene expression in *Salmonella enterica* decrease under anaerobic conditions. *Int. J. Food Microbiol.* **2016**, *238*, 63–67. [CrossRef] [PubMed]

12. Wang, H.; Dong, Y.; Wang, G.; Xu, X.; Zhou, G. Effect of growth media on gene expression levels in *Salmonella* Typhimurium biofilm formed on stainless steel surface. *Food Control* **2016**, *59*, 546–552. [CrossRef]

13. Wang, H.; Zhang, X.; Dong, Y.; Xu, X.; Zhou, G. Insights into the transcriptome profile of mature biofilm of *Salmonella* Typhimurium on stainless steels surface. *Food Res. Int.* **2015**, *77*, 378–384. [CrossRef]

14. Li, J.; Feng, J.; Ma, L.; de la Fuente Núñez, C.; Gölz, G.; Lu, X. Effects of meat juice on biofilm formation of Campylobacter and Salmonella. *Int. J. Food Microbiol.* **2017**, *253*, 20–28. [CrossRef] [PubMed]

15. Lamas, A.; Fernandez-No, I.C.; Miranda, J.M.; Vázquez, B.; Cepeda, A.; Franco, C.M. Prevalence, molecular characterization and antimicrobial resistance of *Salmonella* serovars isolated from northwestern Spanish broiler flocks (2011–2015). *Poult. Sci.* **2016**, *95*, 2097–2105. [CrossRef] [PubMed]

16. Stepanović, S.; Ćirković, I.; Ranin, L. Biofilm formation by *Salmonella* spp. and Listeria monocytogenes on plastic surface. *Lett. Appl. Microbiol.* **2004**, *38*, 428–432. [CrossRef] [PubMed]

17. Birk, T.; Ingmer, H.; Andersen, M.; Jørgensen, K.; Brøndsted, L. Chicken juice, a food-based model system suitable to study survival of *Campylobacter jejuni*. *Lett. Appl. Microbiol.* **2004**, *38*, 66–71. [CrossRef] [PubMed]

18. Römling, U.; Bokranz, W.; Rabsch, W.; Zogaj, X.; Nimtz, M.; Tschäpe, H. Occurrence and regulation of the multicellular morphotype in *Salmonella* serovars important in human disease. *Int. J. Med. Microbiol.* **2003**, *293*, 273–285. [CrossRef] [PubMed]

19. Karatzas, K.A.G.; Randall, L.P.; Webber, M.; Piddock, L.J.V.; Humphrey, T.J.; Woodward, M.J.; Coldham, N.G. Phenotypic and proteomic characterization of multiply antibiotic-resistant variants of *Salmonella enterica* serovar typhimurium selected following exposure to disinfectants. *Appl. Environ. Microbiol.* **2008**, *74*, 1508–1516. [CrossRef] [PubMed]

20. White, A.P.; Gibson, D.L.; Grassl, G.A.; Kay, W.W.; Finlay, B.B.; Vallance, B.A.; Surette, M.G. Aggregation via the red, dry, and rough morphotype is not a virulence adaptation in Salmonella enterica serovar Typhimurium. *Infect. Immun.* **2008**, *76*, 1048–1058. [CrossRef] [PubMed]

21. Seixas, R.; Machado, J.; Bernardo, F.; Vilela, C.; Oliveira, M. Biofilm Formation by Salmonella Enterica Serovar 1, 4,[5], 12: i:-Portuguese Isolates: A Phenotypic, Genotypic, and Socio-geographic Analysis. *Curr. Microbiol.* **2014**, *68*, 670–677. [CrossRef] [PubMed]

22. Fabrega, A.; Vila, J. Salmonella enterica serovar Typhimurium skills to succeed in the host: Virulence and regulation. *Clin. Microbiol. Rev.* **2013**, *26*, 308–341. [CrossRef] [PubMed]

23. Kroupitski, Y.; Golberg, D.; Belausov, E.; Pinto, R.; Swartzberg, D.; Granot, D.; Sela, S. Internalization of Salmonella enterica in leaves is induced by light and involves chemotaxis and penetration through open stomata. *Appl. Environ. Microbiol.* **2009**, *75*, 6076–6086. [CrossRef] [PubMed]

24. Lianou, A.; Koutsoumanis, K.P. Strain variability of the biofilm-forming ability of *Salmonella enterica* under various environmental conditions. *Int. J. Food Microbiol.* **2012**, *160*, 171–178. [CrossRef] [PubMed]

25. Genualdi, S.; Nyman, P.; Begley, T. Updated evaluation of the migration of styrene monomer and oligomers from polystyrene food contact materials to foods and food simulants. *Food Addit. Contam. Part A* **2014**, *31*, 723–733. [CrossRef] [PubMed]

26. Castelijn, G.A.A.; van der Veen, S.; Zwietering, M.H.; Moezelaar, R.; Abee, T. Diversity in biofilm formation and production of curli fimbriae and cellulose of Salmonella Typhimurium strains of different origin in high and low nutrient medium. *Biofouling* **2012**, *28*, 51–63. [CrossRef] [PubMed]

27. Koukkidis, G.; Haigh, R.; Allcock, N.; Jordan, S.; Freestone, P. Salad leaf juices enhance *Salmonella* growth, colonization of fresh produce, and virulence. *Appl. Environ. Microbiol.* **2017**, *83*, e02416-16. [PubMed]

28. Addis, M.F.; Tanca, A.; Uzzau, S.; Oikonomou, G.; Bicalho, R.C.; Moroni, P. The bovine milk microbiota: Insights and perspectives from-omics studies. *Mol. Biosyst.* **2016**, *12*, 2359–2372. [CrossRef] [PubMed]

29. Weiler, C.; Ifland, A.; Naumann, A.; Kleta, S.; Noll, M. Incorporation of Listeria monocytogenes strains in raw milk biofilms. *Int. J. Food Microbiol.* **2013**, *161*, 61–68. [CrossRef] [PubMed]

30. Hamadi, F.; Asserne, F.; Elabed, S.; Bensouda, S.; Mabrouki, M.; Latrache, H. Adhesion of Staphylococcus aureus on stainless steel treated with three types of milk. *Food Control* **2014**, *38*, 104–108. [CrossRef]

31. Jindal, S.; Anand, S.; Huang, K.; Goddard, J.; Metzger, L.; Amamcharla, J. Evaluation of modified stainless steel surfaces targeted to reduce biofilm formation by common milk sporeformers. *J. Dairy Sci.* **2016**, *99*, 9502–9513. [CrossRef] [PubMed]

32. Giaouris, E.; Heir, E.; Desvaux, M.; Hébraud, M.; Møretrø, T.; Langsrud, S.; Doulgeraki, A.; Nychas, G.-J.; Kačániová, M.; Czaczyk, K.; et al. Intra- and inter-species interactions within biofilms of important foodborne bacterial pathogens. *Front. Microbiol.* **2015**, *6*, 841. [CrossRef] [PubMed]

33. Almeida, F.A.; Pimentel-Filho, N.J.; Pinto, U.M.; Mantovani, H.C.; Oliveira, L.L.; Vanetti, M.C.D. Acyl homoserine lactone-based quorum sensing stimulates biofilm formation by Salmonella Enteritidis in anaerobic conditions. *Arch. Microbiol.* **2017**, *199*, 475–486. [CrossRef] [PubMed]

34. Proctor, M.E.; Hamacher, M.; Tortorello, M.L.; Archer, J.R.; Davis, J.P. Multistate outbreak of *Salmonella* serovar Muenchen infections associated with alfalfa sprouts grown from seeds pretreated with calcium hypochlorite. *J. Clin. Microbiol.* **2001**, *39*, 3461–3465. [CrossRef] [PubMed]

35. Vestrheim, D.; Lange, H.; Nygård, K.; Borgen, K.; Wester, A.; Kvarme, M.; Vold, L. Are ready-to-eat salads ready to eat? An outbreak of *Salmonella* Coeln linked to imported, mixed, pre-washed and bagged salad, Norway, November 2013. *Epidemiol. Infect.* **2016**, *144*, 1756–1760. [CrossRef] [PubMed]

36. Sivapalasingam, S.; Barrett, E.; Kimura, A.; Van Duyne, S.; De Witt, W.; Ying, M.; Frisch, A.; Phan, Q.; Gould, E.; Shillam, P. A multistate outbreak of *Salmonella enterica* Serotype Newport infection linked to mango consumption: Impact of water-dip disinfestation technology. *Clin. Infect. Dis.* **2003**, *37*, 1585–1590. [CrossRef] [PubMed]

37. Patel, J.; Singh, M.; Macarisin, D.; Sharma, M.; Shelton, D. Differences in biofilm formation of produce and poultry *Salmonella enterica* isolates and their persistence on spinach plants. *Food Microbiol.* **2013**, *36*, 388–394. [CrossRef] [PubMed]

38. Yaron, S.; Römling, U. Biofilm formation by enteric pathogens and its role in plant colonization and persistence. *Microb. Biotechnol.* **2014**, *7*, 496–516. [CrossRef] [PubMed]

39. Kuda, T.; Shibata, G.; Takahashi, H.; Kimura, B. Effect of quantity of food residues on resistance to desiccation of food-related pathogens adhered to a stainless steel surface. *Food Microbiol.* **2015**, *46*, 234–238. [CrossRef] [PubMed]

Review

Understanding Antimicrobial Resistance (AMR) Profiles of *Salmonella* Biofilm and Planktonic Bacteria Challenged with Disinfectants Commonly Used During Poultry Processing

Myrna Cadena [1], Todd Kelman [1], Maria L. Marco [2] and Maurice Pitesky [1,*]

[1]   UC Davis School of Veterinary Medicine, Department of Population Health and Reproduction, Cooperative Extension, One Shields Ave, Davis, CA 95616, USA
[2]   UC Davis, Department of Food Science and Technology, One Shields Ave, Davis, CA 95616, USA
*   Correspondence: mepitesky@ucdavis.edu; Tel.: +1-530-752-3215

Received: 27 June 2019; Accepted: 17 July 2019; Published: 22 July 2019

Abstract: Foodborne pathogens such as *Salmonella* that survive cleaning and disinfection during poultry processing are a public health concern because pathogens that survive disinfectants have greater potential to exhibit resistance to antibiotics and disinfectants after their initial disinfectant challenge. While the mechanisms conferring antimicrobial resistance (AMR) after exposure to disinfectants is complex, understanding the effects of disinfectants on *Salmonella* in both their planktonic and biofilm states is becoming increasingly important, as AMR and disinfectant tolerant bacteria are becoming more prevalent in the food chain. This review examines the modes of action of various types of disinfectants commonly used during poultry processing (quaternary ammonium, organic acids, chlorine, alkaline detergents) and the mechanisms that may confer tolerance to disinfectants and cross-protection to antibiotics. The goal of this review article is to characterize the AMR profiles of *Salmonella* in both their planktonic and biofilm state that have been challenged with hexadecylpyridinium chloride (HDP), peracetic acid (PAA), sodium hypochlorite (SHY) and trisodium phosphate (TSP) in order to understand the risk of these disinfectants inducing AMR in surviving bacteria that may enter the food chain.

Keywords: *Salmonella*; biofilm; disinfectants; poultry; transcriptome; resistance

1. Introduction

Salmonella is a major foodborne pathogen worldwide and is highly associated with contaminated poultry products. In the United States alone, the Centers for Disease Control and Prevention (CDC) estimates that *Salmonella* causes approximately 1.2 million foodborne illnesses, 23,000 hospitalizations and 450 deaths per year [1]. In addition to causing foodborne illness, *Salmonella* isolates from poultry products and processing plants have been found to be both tolerant to disinfectants and resistant to antibiotics despite not being challenged with antibiotics during poultry production and/or processing [2–5]. Furthermore, studies have shown positive correlations between tolerance to disinfectants and resistance to antibiotics in poultry products [4,5]. Growing concerns over disinfectants conferring cross-protection to antibiotics has increased focus on understanding the mechanisms of antimicrobial resistance (AMR) in bacteria, and specifically foodborne pathogens [6]. The present review aims to understand how commonly used disinfectants such as hexadecylpyridinium chloride (HDP), peracetic acid (PAA), sodium hypochlorite (SHY) and trisodium phosphate (TSP) may confer AMR in *Salmonella* in order to evaluate the risk of these disinfectants increasing AMR in surviving *Salmonella*.

Biofilms are organized structures of bacterial cells that produce a self-encasing polymer extracellular matrix and can adhere to biotic (living) and abiotic (inert/nonliving) surfaces [7].

As opposed to the more commonly studied and well understood planktonic (free-floating) form, biofilms are the predominant form of bacterial growth. It is estimated that 80% of all infections in humans are thought to be of biofilm origin [8]. Various abiotic substrates, such as Teflon™, stainless steel, rubber and polyurethane can support *Salmonella* biofilm adherence and growth [9,10], which are regulated by various environmental conditions such as pH, temperature and NaCl concentration [11]. Within a poultry processing facility setting, *Salmonella* and *Campylobacter* biofilm formation is facilitated by the presence of meat juice on abiotic surfaces under static and flow conditions [12]. Formation of biofilms provides ecologic advantages to the enclosed bacteria, including protection from the environment (e.g., temperature, pH and osmotic extremes, UV light exposure, desiccation), increased nutrient availability, metabolic enhancement, and facilitation of gene transfer [13]. Additionally, biofilm formation confers increased antimicrobial resistance through a variety of mechanisms. From a practical perspective, biofilm cells are 10- to 1000-fold less susceptible to anti-microbial agents than the planktonic form of the same bacterium [8,14,15]. In poultry processing plants, the use of sub-optimal concentrations of a commonly applied biocide (peracetic acid) has been demonstrated to facilitate the persistence of *Salmonella* biofilms [16]. All of these characteristics appear to be facilitated at least in part by an intercellular communication mechanism known as quorum sensing—small signal molecules called autoinducers exchanged between bacteria as a function of population density. These signal molecules can regulate expression of numerous genes, including those associated with biofilm adherence, metabolism, and virulence. The development of inhibitors of such factors may be key to controlling biofilm formation and pathogenicity [10,13,17–19].

2. Disinfectants Commonly Used during Poultry Processing

In the most general sense, poultry processing can be divided into two phases—first processing and second processing (see Figure 1) [20]. First processing consists of carcass receiving to chilling. This step includes scalding, defeathering and evisceration. Second processing encompasses carcass chilling to shipping. This step includes packaging and may include carcass/parts processing.

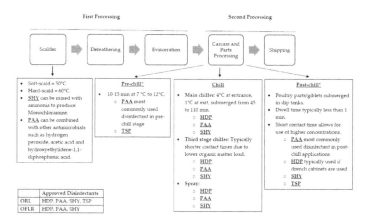

Figure 1. A general schematic overview of commercial poultry processing. The information provided serves as an example of possible scenarios. However, protocols can vary widely among processing plants. In addition, the list of antimicrobials approved for on-line reprocessing (OLR) and off-line reprocessing (OFLR) is dynamic in terms of application and concentration. * At a minimum, sampling for pathogens must occur at the pre- and post-chill points.

Moreover, during processing, carcasses that are contaminated with fecal matter or digestive tract content as determined by visual inspection right after evisceration may still pass inspection if reprocessed properly [21]. Two methods of reprocessing include on-line reprocessing (OLR) and off-line reprocessing (OFLR). Processing plants may use either or both forms of reprocessing. In OLR, contaminated carcasses can be reprocessed manually by trimming away contaminated parts and/or treated with an antimicrobial along with poultry carcasses that are not contaminated on the main line. In contrast, OFLR entails taking out contaminated carcasses off the main line where they can be reprocessed manually by trimming away contaminated parts and/or treated with an antimicrobial, away from the visually uncontaminated carcasses. Processing plants may use either or both forms of reprocessing. Then, before carcass chilling, carcasses are visually inspected again for fecal contamination using guidelines provided by the Food Safety Inspection and Inspection Service Directive in order to comply with the zero-tolerance standard that requires carcasses to be free of fecal contamination before entering the chiller tank [22,23]. While disinfectants are used in both first and second processing, their application, contact time and temperature may differ. Similarly, disinfectants approved for use during OLR and OFLR may differ in the application, contact time and temperature [24]. For the purposes of this review, only disinfectants commonly used during poultry processing will be reviewed, as intervention protocols may vary greatly across poultry processing plants. Popular disinfectants include hexadecylpyridinium chloride (HDP), peracetic acid (PAA), sodium hypochlorite (SHY) and trisodium phosphate (TSP) [25,26]. While PAA is the most commonly used disinfectant in both pre- and post-chill applications, HDP is most commonly used in post-chill applications when drench cabinets are used [27]. Additionally, both HDP and PAA can be used for OLR and OFLR [24]. SHY and TSP are also commonly used in post-chill applications [25]. In terms of reprocessing, SHY is only allowed in OFLR, although when it is used in combination with other disinfectants it can be used in OLR as well. In contrast, TSP is only allowed in OLR. Figure 1 provides a schematic overview of poultry processing along with information on HDP, PAA, SHY and TSP.

3. Poultry Processing Methods Conferring Biocide Tolerance

Currently, studies suggest that biocide tolerance provides cross-protection to various antimicrobials, including antibiotics [28]. Both repeated exposure and sub-inhibitory concentrations of biocides have been shown to allow bacteria to adapt to biocides resulting in biocide-tolerant, antibiotic-resistance bacteria [28–30]. As seen in Figure 1, pathogens such as *Salmonella* may undergo repeated exposure to disinfectants during poultry processing especially if they are allowed in OLR and OFLR. In addition, challenging *Salmonella* with sub-inhibitory concentrations of disinfectants at the processing plant can occur due to the presence of high loads of organic material, inadequate distribution, high prevalence of biofilms, inadequate mixing, preparation, and concentration of biocides. Chicken carcasses have high amounts of organic material that can inactivate certain classes of disinfectants, including quaternary ammonium compounds (HDP) and halogens (SHY) [31]. Additionally, *Salmonella* can adhere to areas that are not easily accessible to disinfectants, in places such as crevices and feather follicles on poultry skin [31,32].

With respect to biofilms, the extracellular matrix limits access to stressors, thereby providing some protection against disinfectants and antibiotics that allow *Salmonella* to persist on biotic (e.g., live birds) and abiotic (e.g., stainless steel) surfaces [6]. While biofilms are more prevalent than the easier-to-kill planktonic form in processing plants, the common methodology for evaluating the efficacy of disinfectants requires identifying the inhibitory concentration and or log reduction of the bacteria in their planktonic form [33]. Therefore, if the tests used to determine effective disinfectant concentrations are done on planktonic bacteria, it is possible that the concentrations are sub-inhibitory in the processing plant. Shah et al. demonstrated that when *Salmonella* Typhimurium was preadapted to cold stress, it was tolerant to subsequent acid stress [34]. Therefore, it is possible that *Salmonella* passing the chilling stage may also be harder to kill and require a higher concentration of disinfectant at the post-chill tank.

Overall, processing methods may be priming bacteria to stressors found throughout the slaughtering process resulting in surviving bacteria that are tolerant to biocides and resistant to antibiotics. Therefore, it is becoming increasingly important to understand how methods of pathogen control in food processing can be improved in terms of reducing tolerance to disinfectants and resistance to antibiotics. The specific mechanisms that can confer biocide tolerance and antibiotic resistance in bacteria exposed HDP, PAA, SHY and TSP will be discussed next.

4. Proposed Mechanisms of Bacterial Resistance to Antimicrobials Induced by Disinfectants Used During Poultry Processing

Cross-resistance, or resistance to a variety of substances via a physiological adaptation, as opposed to genetic linkage as is the case with co-resistance, is an important mechanism of bacterial resistance to antimicrobials [28]. Examples of cross-resistance mechanisms include reduced cell permeability, production of neutralizing enzymes, target alteration and overactive efflux pumps which can pump out a broad-spectrum of substances including antibiotics, biocides and other inhibitors out of the cell and create multidrug resistant (MDR) bacteria [6,35].

In particular, overactive efflux pumps and changes to the outer membrane have been proposed as broad-spectrum mechanisms conferring tolerance and/or resistance to antimicrobials in *Salmonella* after exposure to HDP and TSP [4,36] (Table 1). To evaluate these proposed mechanisms, Mavri et al. exposed TSP- and HDP- adapted *Campylobacter jejuni* and *Campylobacter coli* (Gram-negative bacteria like *Salmonella*, which share a superfamily of efflux pumps known as Resistance-Nodulation-Division [37,38]) to efflux pump inhibitors and evaluated their outer membrane proteins [30]. Results showed that TSP- adapted *Campylobacter jejuni* and *Campylobacter coli* had a weaker adaptive resistance to TSP and weak cross-resistance to antibiotics compared to HDP-adapted *Campylobacter jejuni* and *Campylobacter coli* [30]. The authors proposed that different efflux systems play a role in cross-resistance due to different modes of action of the disinfectants resulting in different levels of cross-resistance [30]. Additionally, TSP caused greater reduction in outer membrane protein (OMP) content than HDP, resulting in the most damaging effects on bacterial cells [30]. It was also noted that some strains of *Campylobacter* displayed increased susceptibility to biocides after repeated exposure—Mavri et al. proposed that some cell envelope modifications may actually promote biocide uptake [30]. Along with linking efflux pumps to cross-resistance, this study revealed that mechanisms involved in biocide (e.g., triclosan, benzalkoniumchloride, hexadecylpyridinium chloride, chlorhexidine diacetate and trisodium phosphate) adaptation are unique for various strains of *Campylobacter* as opposed to it having only one species-specific mechanism [30]. These findings suggest that utilizing a serotype-specific or even a strain-specific approach to select disinfectants is becoming increasingly important.

In addition to AMR and MDR, efflux pumps have also been associated with increased invasion in *Salmonella* [43]. For example, in addition to regulating resistance to fluroquinolones in *Salmonella*, the acrAB operon—part of the AcrAB-Tolc multidrug efflux pump—has also been shown to be upregulated during sub-inhibitory exposure to the bile salt sodium deoxycholate (DOC), particularly during exponential growth [43,44]. DOC at high concentrations exhibits biocidal-like activity including disruption of cell membranes, denaturation of proteins and oxidative DNA damage [44]. By adapting to DOC, *Salmonella* Typhimurium can then proliferate and continue to invade the host, while strains lacking AcrAB-Tolc were unable to adapt to DOC [44]. From a food safety perspective, cross-resistance imposes a food safety hazard in that repeated exposure to biocides can potentially induce biocide tolerance, AMR and increased virulence in bacteria entering the food chain [28].

Table 1. Mode of action (MOA) and proposed mechanism of antibiotic resistance for the following commonly used disinfectants in poultry processing: HDP, PAA, SHY and TSP (HDP: hexadecylpyridinium chloride, PAA: peracetic acid, SHY: sodium hypochlorite, TSP: trisodium phosphate).

Disinfectant	Disinfectant Type	Proposed Modes of Action	References	Proposed Mechanism Conferring Antibiotic Resistance	References
HDP	Quaternary ammonium	• Adsorption and penetration of cell wall. • Disruption of cytoplasmic membrane. • Leakage of intracellular low molecular-weight constituents. • Degradation of proteins and nucleic acids. • Cell lysis due to cell wall degrading autolytic enzymes.	[30,39]	• Overexpression of efflux pumps. • Induce cellular morphological changes such as thickening of cell envelope or loss in outer membrane proteins.	[39]
PAA	Organic acid and an oxidant	• Non-specific oxidation particularly of C–C double bonds and reduced atoms (i.e., S).	[40]	• None known.	[40]
SHY	Chlorine	• Uncoupling of the electron chain or enzyme inactivation (i.e., trans-phosphorylase inactivation) either in the membrane or in the cell interior.	[41]	• Induces biofilm formation.	[42]
TSP	Alkaline detergent	• High pH (12 to 13) disrupts cytoplasmic and outer membranes resulting in leakage and eventual cell death. • High ionic strength can cause bacterial cell autolysis. • Removes bacterial cells from carcass surface (i.e., chicken skin) by removing a thin layer of lipids ("detergent" effect) from the surface of the carcass thereby exposing cells that would otherwise be protected, and results in bacterial cell autolysis.	[30,41]	• Overexpression of efflux pumps. • Induce cellular morphological changes such as loss in outer membrane proteins.	[30]

HDP: hexadecylpyridinium chloride, PAA: peracetic acid, SHY: sodium hypochlorite, TSP: trisodium phosphate.

In contrast to the increased susceptibility of *Campylobacter* to biocides after exposure, SHY has been shown to induce biofilm production in *Pseudomonas aeruginosa* (Table 1), also a Gram-negative bacteria like *Salmonella* [42]. As discussed previously, biocide tolerance and antibiotic resistance can be attributed to biofilms, as the extracellular matrix provides the cells protection against disinfectants and antibiotics [6], while the clustering of cells may facilitate the transfer of antimicrobial resistance genes via horizontal gene transfer [45]. Extracellular DNA may also play a role in the proliferation of biofilms in *Salmonella* and other bacteria: *Staphylococcus epidermidis* biofilm has been shown to have a strong binding affinity to vancomycin thereby limiting access to cells [46,47]. Additionally, RNA-sequencing analysis of planktonic and biofilm *Salmonella*, indicates that gene expression patterns differ between the two forms under the same acid stress [48,49]. Furthermore, RNA-sequencing suggests that in *Salmonella* Typhimurium the same environmental stressors results in upregulation of virulence genes in the planktonic form—priming that population for host invasion rather than for environmental survival as it does for the biofilm counterpart [50]. More studies that investigate the transcriptome or resistome of bacteria challenged with disinfectants are needed. From a Hazard Analysis and Critical Control Points (HACCP) perspective, utilizing RNA-sequencing could be used to determine critical food safety parameters in a food system environment with the ultimate goal of identifying conditions in food production that mitigate transcription of genes associated with AMR and virulence. From a practical perspective, integrating Whole Genome Sequencing (WGS) and RNA-seq of selected isolates collected during routine surveillance in the processing facility could be used as a way to validate and optimize disinfectant selection.

Unlike HDP, SHY and TSP, PAA does not have a proposed mechanism for conferring antibiotic resistance or even tolerance (Table 1). Because PAA has two distinct modes of action due to being an organic acid and an oxidant, it is theorized that a cell is less likely to develop tolerance or resistance mechanisms against PAA or antibiotics [40]. This provides valuable information in that in addition to being effective for the control of both *Salmonella* and *Campylobacter* [51], PAA also seems like it is less likely to induce AMR and may even decrease it. One approach could be to utilize PAA at the last step of cleaning and disinfection with the goal of reducing incidence of AMR.

5. Antimicrobial Resistance Profiles of Foodborne Pathogens Challenged with Disinfectants

Table 2 provides AMR profiles for HDP, PAA, SHY and TSP, which demonstrates that biocides can differ in the way they induce AMR across different organisms. PAA-challenged *E. coli* resulted in an overall decrease in antimicrobial resistance gene classes [52]. This is in line with the theory that PAA is less likely to induce AMR. HDP-challenged *Salmonella* strains showed a decrease, an increase or both in resistance to certain antibiotics after repeated exposure relative to the wildtype, resulting in mixed effects (Table 2). This emphasizes the importance of testing disinfectants with different serovars and not just single strains of a species as described for *Campylobacter* [30]. One study by Molina-Gonzalez et al. suggests that SHY and TSP can induce AMR when *Salmonella* Enteritidis, *Salmonella* Kentucky and *Salmonella* Typhimurium are exposed to those disinfectants at sub-inhibitory concentrations [36]. Although the experiments from Table 2 cannot be compared directly due to differences in experimental design and analysis, they all provide evidence in support of the conclusion that that proper utilization of disinfectants should include consideration of those biocides that are less likely to increase AMR and biocide tolerance. Therefore, utilizing a serotype-specific approach when selecting disinfectants should be considered by poultry processing facilities.

Table 2. AMR profiles of isolates challenged with disinfectants commonly used during poultry processing.

Reference	Disinfectant	Application Parameters	Isolate	AMR Profile		
				Showed increased resistance (i.e., twice the MIC) to the following antimicrobials compared to the wildtype:		**Showed decreased resistance to the following antimicrobials compared to the wildtype:**
[29]	HDP	Exposed to increasing concentrations of CPC (0.01, 0.1, 1, 5, 10, 50, 100, 200, 500 mg/mL, 1, 2, 5 and 10 mg/mL).	Salmonella UJA591	Ampicillin, Sulfamethoxazole, Nalidixic acid		Ceftazidime
			Salmonella UJA82k			Ceftazidime
			Salmonella UJA82l	Nalidixic acid		Ampicillin, Cefotaxime, Ceftazidime, Sulfametoxazol
				HDP tolerance level:		**Showed resistance to:**
[4]	HDP	HDP tolerance and antibiotic resistance were determined by using MIC assays.	Salmonella spp. UJAS6	Tolerant		Ampicillin, Chloramphenicol, Tetracycline, Nalidixic acid, Trimethoprim-sulfamethoxazole
			S. enterica UJAS10	Tolerant		Ampicillin, Tetracycline, Nalidixic acid, Trimethoprim-sulfamethoxazole
			Salmonella spp. UJAS18	Tolerant		Ampicillin, Cefotaxime; Ceftazidime, Ciprofloxacin, Chloramphenicol, Tetracycline, Netilmicin, Nalidixic acid, Trimethoprim-sulfamethoxazole
			Salmonella spp. UJAS19	Tolerant		Cefotaxime; Ceftazidime, Ciprofloxacin, Chloramphenicol, Streptomycin, Tetracycline, Netilmicin, Nalidixic acid, Trimethoprim-sulfamethoxazole
				MIC fold change of Campylobacter strains relative to the pre-adapted strains.		
[30]	HDP	Step-wise exposure to gradually increasing concentrations (2, 2.5, 3, 4 to 5 mg/mL, depending upon the growth of the adapted microorganism) of HDP over 15 days.	Campylobacter jejuni K49/4	Days after repeated exposure to HDP	5	10	15
			Campylobacter jejuni NCTC11168	MIC fold change	1	1	1
				Days after repeated exposure to HDP	5	10	15
			Campylobacter jejuni ATCC33560	MIC fold change	2	1	4
				Days after repeated exposure to HDP	5	10	15
			Campylobacter coli 137	MIC fold change	0.5	1	1
				Days after repeated exposure to HDP:	5	10	15
			Campylobacter coli ATCC33559	MIC fold change	1	1	0.5
				Days after repeated exposure to HDP	5	10	15
				MIC fold change	1	2	2
				Number of antimicrobial resistance gene classes in PAA treated strains:		
[52]	PAA	Exposed to 0.9 to 2.0 mg/L of PAA to reach target disinfection level of 200 CFU/100mL	Escherichia coli	Mean number of classes decreased by an average of 47% with significant reductions in the following classes: Macrolides (−62.3%), Beta-lactams (−41.3), Phenicols (−64) and Trimethoprim (−49.9).		

Table 2. *Cont.*

Reference	Disinfectant	Application Parameters	Isolate	AMR Profile
				Showed increased resistance (i.e., susceptible to resistant via disk diffusion assay) after exposure to disinfectant:
[36]	SHY	Exposed to increasing sub-inhibitory concentrations (starting at MIC/2).	Salmonella Enteritidis	Ceftazidime
			Salmonella Kentucky	Amikacin, Ampicillin/ sulbactam
			Salmonella Typhimurium	Amikacin, Tobramycin, Cefazolin, Cefotaxime
	TSP		Salmonella Enteritidis	Amikacin, Cefazolin, Cefoxitin, Ceftazidime, Aztreonam, Nalidixic acid, Phosphomycin
			Salmonella Kentucky	Amikacin, Ceftazidime, Aztreonam, Phosphomycin
			Salmonella Typhimurium	Amikacin, Cephalothin, Cefazolin, Cefoxitin, Cefepime, Aztreonam, Phosphomycin
				Mean number of antibiotics the strains were resistant to at 0 days of storage:	**Mean number of antibiotics the strains were resistant to after 5 days of storage:**
[31]	TSP	Chicken legs containing *E. coli* were dipped in 12% TSP at 20 ± 1 °C for 15 min and subsequently refrigerated at 7 ± 1 °C and stored. Chicken legs dipped in tap water were used as a control.	*Escherichia coli*	Control: 3.76 ± 2.01 [a] [a] TSP: 3.80 ± 2.48 [a] [a]	Control: 3.44 ± 1.42 [a] [a] TSP: 4.64 ± 2.64 [b] [b]

The mean numbers from the same day (different treatments) with no letters in common (superscript) are significantly different ($P < 0.05$). The mean numbers within the same treatment (day 0 versus day 5) with no letters in common (subscript) are significantly different ($P < 0.05$).

MIC fold change of *Campylobacter* strains relative to the pre-adapted strains.

Reference	Disinfectant	Application Parameters	Isolate			
[30]	TSP	Step-wise exposure to gradually increasing concentrations (2, 2.5, 3, 4 to 5 mg/mL, depending upon thegrowth of the adapted microorganism) of TSP over 15 days.	*Campylobacter jejuni* K49/4	Days after repeated exposure to TSP	5	10	15
				MIC fold change	2	2	2
			Campylobacter jejuni NCTC11168	Days after repeated exposure to TSP	5	10	15
				MIC fold change	2	0.5	2
			Campylobacter jejuni ATCC33560	Days after repeated exposure to TSP	5	10	15
				MIC fold change	1	1	0.125
			Campylobacter coli 137	Days after repeated exposure to TSP	5	10	15
				MIC fold change	1	0.008	0.004
			Campylobacter coli ATCC33559	Days after repeated exposure to TSP	5	10	15
				MIC fold change	2	2	1

HDP: hexadecyzpyridinium chloride; PAA: peracetic acid; SHY: sodium hypochlorite; TSP: trisodium phosphate.

6. Biofilm Detection

Generally, biofilm-producing strains have been identified quantitatively by microtiter-plate assays or qualitatively by the Congo red agar or test tube methods, both of which use a phenotypic approach [53,54]. Genotypic identification of biofilm-producing strains relies on molecular methods to detect biofilm-associated genes by conventional PCR, qPCR or multiplex PCR [10,55]. The *csgD* gene in *Salmonella* Typhimurium has been identified as a central biofilm regulator gene in which bistable expression allows for either increased virulence or persistence in the environment [50]. Additionally, genes associated with curli, fimbriae, cellulose such as *csgD*, *csgB*, *adrA*, and *bapA* have been utilized to detect *Salmonella* biofilms on eggshells. Furthermore, genes related to flagella adhesion, metabolism, regulation/stress response and proteic envelop/secretion can be used to classify biofilm formation capacity and flagellar motility [18].

By using a broad set of phenotypic and genotypic techniques such as the ones mentioned above, it is now well understood that *Salmonella* biofilms are associated with persistence both inside and outside the host including on poultry carcasses and processing plants even after cleaning and disinfection [7,56]. Sensory inspections of open surfaces such as visual, tactile and olfactory observations such as greasy surfaces allow for quick identification of obvious issues in the sanitation process. However, it is important to note that bacterial counts are not correlated with visual inspections [57]. Additionally, while food contact surfaces have become well-established sources of contamination and recontamination in food processing settings [58,59], Arnold and Silvers [60] found that microbial attachment and biofilm formation vary depending on surface type (e.g., stainless steel, conveyor belting, polyethylene and picker-finger rubber). Interestingly, contrary to previous studies that examined planktonic bacteria, picker-finger rubber commonly used in defeathering machines were shown to inhibit microbial contamination and biofilm formation [60]. However, more testing on different combinations of strains observed at the processing plant need to be conducted since it has been shown that microbial attachment and biofilm properties may behave differently depending on the combination of strains that make up the bacterial community [61]. Therefore, based on these considerations, careful and robust assessments of open surfaces at the processing plant should be considered even when it has been well established that *Salmonella* and *Campylobacter* have been isolated from poultry production and processing [9,57].

In summary, poultry processing plants should consider taking measures to detect and characterize biofilms from their specific facility to optimize the prevention and management of biofilms. Fortunately, there are now direct and indirect approaches that can be applied at the food processing plant to detect the presence of biofilms through direct observation on open surfaces and to quantify cells isolated from biofilms found at the food processing plant. Direct methods directly observe biofilm colonization, whereas indirect methods start with detaching biofilm from food-contact surfaces before quantifying them [57]. Commercially available and easy-to-use tools that detect the presence of biofilms include BioFinder (Barcelona, Spain) [62], REALCO Biofilm Detection Kit (Louvain-la-Neuve, Belgium) [63], TBF® 300 (Valencia, Spain) and TBF® 300S [64]. TEMPO® system (Marcy l'Etoile, France) allows for quantification via the most probable number (MPN) technique and is also commercially available [65]. Table 3 summarizes biofilm detection methods used in food processing settings. These tools could help improve the eradication of biofilms and can be used to evaluate current cleaning procedures [57]. At the same time, lack of consensus across detection methods should be taken into consideration and a combination of methods for detection and enumeration may need to be implemented [66].

Table 3. Direct and indirect biofilm detection and enumeration methods for food processing settings.

Test	Type	Method	References
Direct			
BioFinder	Qualitative	Direct observation of color change due to dying of biofilm components.	[62]
Contact plates	Quantitative	Sterile agar plate is placed on surface of interest and biofilm is detected via conventional culture methods.	[67]
Direct epifluorescence microscopy	Quantitative	Automatic cell quantification using computer software on digital images.	[68]
REALCO Biofilm Detection Kit	Qualitative	Direct observation of color change due to dying of biofilm components.	[63]
TBF® 300/ TBF® 300S	Qualitative	Direct observation of color change due to dying of biofilm components.	[64]
Indirect			
BacTrac 4300	Quantitative	Total viable counts calculated via impedance.	[69,70]
Plate count	Quantitative	Culture plating to determine the number of colony forming units (CFU).	[57]
TEMPO®	Quantitative	Cell counts from biofilms are calculated using most probable number (MPN) system based on fluorescence.	[65]
Abcam XTT tetrazolium salt and resazurin assay kit	Quantitative	Metabolic assays combined with spectrophotometry can be used to quantify biofilm.	[57,71,72]

7. Biofilm Characterization

In addition to taking measures to detect biofilm-producing bacteria, another important step for processing plants to consider is the characterization of biofilms present at the processing plant. Biofilm characterization can help improve food safety as several differences in biocide susceptibility, pathogenicity and persistence in the environment between biofilms and planktonic bacteria have been identified.

Minimum inhibitory concentration (MIC) and minimum biofilm eliminating concentrations (MBEC) assays have traditionally been used to determine the efficacy of antibiotics on planktonic and biofilm bacteria, respectively [73]. However, these assays can also be used to determine the efficacy of other biocides such as disinfectants [33]. Furthermore, results from these assays can be used to directly compare planktonic and biofilm bacterial forms. As an example, Chylkova, Cadena, Ferreiro and Pitesky [33] found that acidified calcium hypochlorite (aCH) and PAA were ineffective against *Salmonella* biofilms at contact and concentrations commonly used during poultry processing whereas HDP remained effective, based on MIC and MBEC assays. Similarly, PAA has been found to be inefficient at eliminating *Salmonella* biofilms from polypropylene and polyurethane surfaces which are common surface types used in poultry processing plants [16]. Sarjit and Dykes [74] found that trisodium phosphate (TSP), unlike sodium hypochlorite (SHY), was an effective sanitizer against biofilms on stainless steel, glass and polyurethane surfaces. In contrast, Korber et al. [75] observed *Salmonella* biofilm cells from glass surfaces were less susceptible to TSP. Differences in biofilm response to disinfectants and surfaces indicate further testing is necessary to further elucidate biofilm formation at the processing plant.

In addition to phenotypic differences as shown by differences in biocide susceptibility, genotypic differences between planktonic and biofilm bacterial cells have also been shown. Wang et al. [76] found that planktonic and biofilm Salmonella Typhimurium cells isolated from raw chicken meat and contact surfaces from poultry processing plants showed distinct gene expression patterns. Specifically, genes from gene ontology groups related to membrane proteins, cytoplasmic proteins, curli productions, transcriptional regulators, cellulose biosynthesis and stress response proteins were differentially expressed suggesting they may play a role in biofilm maturation. Furthermore, virulence and persistence genes have been shown to be differentially expressed in planktonic and biofilm *Salmonella* Typhimurium cells with planktonic cells expressing genes associated with virulence and biofilms expressing genes associated with environmental persistence [50]. These results are in line with results from Borges et al. [77] in which *Salmonella* Typhimurium biofilm production was not associated with in vivo pathogenicity index (PI). However, there was an association between biofilm formation and PI in *Salmonella* Enteritidis at 28 °C [77]. Interestingly, in microaerobiosis and anaerobiosis conditions, *Salmonella* Typhimurium grown in chicken residue displayed downregulation of biofilm associated genes (e.g., *csgD* and *adrA*) and upregulation of virulence genes (e.g., *hilA* and *invA*) on stainless steel [78]. Contrastingly, biofilm formation was upregulated in aerobiosis [78]. Results showed that oxygen levels could have an effect on biofilm formation. Similarly, a study by Wang et al. [79] showed that growth media could also have an effect on biofilm gene expression with biofilm grown in laboratory trypticase soy broth (TSB) expressing upregulation of biofilm formation genes and biofilm grown on meat thawing loss broth (MTLB) expressing inhibited gene expression.

It is well known that there is a correlation between *Salmonella* biofilm formation and persistence in factory environments; however, several studies have shown that different serovars of *Samonella* have different capacities to make biofilm under different environmental conditions [80–82]. For example, when studying *Salmonella* Enteritidis, Infantis, Kentucky and Telaviv serotypes at different temperatures, a shift in biofilm formation capacities was observed with most of the serotypes becoming strong biofilm producers at 22 °C [82]. In contrast, at 37 °C, only some of the *Salmonella* Enteritidis and Infantis serovars were considered strong biofilm producers. In addition to temperature, pH and NaCl concentrations have been shown affect *Salmonella* strains ability to form biofilm [80]. Conditions that were unfavorable and increased biofilm formation in most of the *Salmonella* strains were pH 5.5, 0.5% NaCl and 25 °C [80]. Surface materials also lead to differences in microbial adhesion with polyurethane displaying more irregular adhesion than polypropylene [16]. Moreover, no viable cells were isolated from polypropyle after treatment with sanitizers commonly used in Brazilian poultry processing plants. Overall, these results show that environmental conditions, growth media and strains can influence Salmonella biofilm formation thus highlighting the importance of mimicking food processing conditions during biofilm and disinfectant efficacy testing.

In summary, the persistence and complexity of *Salmonella* biofilms in food processing plants suggest that protocols used to control and eliminate biofilms should be constantly evaluated and modified accordingly. This is especially relevant when considering that many of the cleaning products used in the food industry are not optimized for the elimination of biofilms [83] and an estimated 65% of human bacterial infections are caused by biofilms [84]. It is also important to note that in vitro efficacy testing of disinfectants should be done on mature biofilms as that is the bacterial form and stage most commonly found in food processing environments particularly food-contact surfaces [85]. Likewise, testing conditions should mimic poultry processing conditions in order to improve the applicability of the results as strain, environmental conditions and growth mediums have been shown to influence biofilm formation. Furthermore, alternative biofilm eradication methods that act specifically on biofilms should also be considered, such as lactic acid bacteria, phagetherapy, crude essential oils, quorum sensing inhibitors and bacteriocins [10,86]. However, it should also be noted that the most effective strategy would be to prevent biofilm formation in the first place [10].

8. Conclusions

While more research is needed to further our understanding of AMR profiles from pathogens isolated from poultry processing facilities, this review suggests that understanding what AMR mechanisms are activated by disinfectants can provide poultry processing facilities insights as to which disinfectants to use at a particular facility. Therefore, active monitoring of pathogens present at the grow-out facility and utilizing that information to strategize which disinfectants to employ at the processing plant (i.e., serotype-specific approach) should be considered. The efficacy of disinfectants on

Foods **2019**, *8*, 275

biofilms in addition to planktonic bacteria should be frequently tested in order to monitor for changes in susceptibility to disinfectants and prevent the use of sub-inhibitory concentrations. Using disinfectants that differ in their modes of action throughout poultry processing is also advisable as this can potentially reduce the ability of the bacteria to adapt and become tolerant to biocides and antimicrobials. The effect of the disinfectants on the transcriptome or resistome of the pathogen in question may be the key to furthering our understanding of AMR. Alternative approaches to the control of planktonic bacteria and biofilm formation that do not rely on the use of traditional biocides, such as enzymes, bacteriophages and quorum sensing inhibitors may be valuable to controlling microbial contamination without inducing AMR, and thus may be the new horizon of antimicrobial food safety [7,9,10,13,87].

Author Contributions: M.C., M.P. and T.K. reviewed the literature and drafted the manuscript. M.C., M.P., T.K and M.L.M. read and revised the manuscript. M.C., M.L.M., T.K. and M.P. responsible for the concept and preparation of final article.

Funding: This research received no external funding.

Conflicts of Interest: The authors declare no conflicts of interest.

References

1. CDC. Salmonella. Available online: https://www.cdc.gov/salmonella/index.html (accessed on 20 July 2019).
2. Youn, S.Y.; Jeong, O.M.; Choi, B.K.; Jung, S.C.; Kang, M.S. Comparison of the antimicrobial and sanitizer resistance of *Salmonella* isolates from chicken slaughter processes in Korea. *J. Food Sci.* **2017**, *82*, 711–717. [CrossRef] [PubMed]
3. Mion, L.; Parizotto, L.; Calasans, M.; Dickel, E.L.; Pilotto, F.; Rodrigues, L.B.; Nascimento, V.P.; Santos, L.R. Effect of antimicrobials on *Salmonella* spp. strains isolated from poultry processing plants. *Braz. J. Poult. Sci.* **2016**, *18*, 337–341. [CrossRef]
4. Marquez, M.L.F.; Burgos, M.J.G.; Pulido, R.P.; Galvez, A.; Lopez, R.L. Biocide tolerance and antibiotic resistance in *Salmonella* isolates from hen eggshells. *Foodborne Pathog. Dis.* **2017**, *14*, 89–95. [CrossRef] [PubMed]
5. Marquez, M.L.F.; Burgos, M.J.G.; Pulido, R.P.; Galvez, A.; Lopez, R.L. Correlations among Resistances to Different Antimicrobial Compounds in Salmonella Strains from Hen Eggshells. *J. Food Prot.* **2018**, *81*, 178–185. [CrossRef] [PubMed]
6. Wales, A.D.; Davies, R.H. Co-selection of resistance to antibiotics, biocides and heavy metals, and its relevance to foodborne pathogens. *Antibiotics* **2015**, *4*, 567–604. [CrossRef] [PubMed]
7. Steenackers, H.; Hermans, K.; Vanderleyden, J.; De Keersmaecker, S.C.J. Salmonella biofilms: An overview on occurrence, structure, regulation and eradication. *Food Res. Int.* **2012**, *45*, 502–531. [CrossRef]
8. Davies, D. Understanding biofilm resistance to antibacterial agents. *Nat. Rev. Drug Discov.* **2003**, *2*, 114–122. [CrossRef] [PubMed]
9. Srey, S.; Jahid, I.K.; Ha, S.-D. Biofilm formation in food industries: A food safety concern. *Food Control* **2013**, *31*, 572–585. [CrossRef]
10. Merino, L.; Procura, F.; Trejo, F.M.; Bueno, D.J.; Golowczyc, M.A. Biofilm formation by Salmonella sp. in the poultry industry: Detection, control and eradication strategies. *Food Res. Int.* **2017**, *119*, 530–540. [CrossRef]
11. Moraes, J.O.; Cruz, E.A.; Souza, E.G.F.; Oliveira, T.C.M.; Alvarenga, V.O.; Peña, W.E.L.; Sant'Ana, A.S.; Magnani, M. Predicting adhesion and biofilm formation boundaries on stainless steel surfaces by five Salmonella enterica strains belonging to different serovars as a function of pH, temperature and NaCl concentration. *Int. J. Food Microbiol.* **2018**, *281*, 90–100. [CrossRef]
12. Li, J.; Feng, J.; Ma, L.; de la Fuente Núñez, C.; Gölz, G.; Lu, X. Effects of meat juice on biofilm formation of Campylobacter and Salmonella. *Int. J. Food Microbiol.* **2017**, *253*, 20–28. [CrossRef] [PubMed]
13. Giaouris, E.E.; Simões, M.V. Pathogenic Biofilm Formation in the Food Industry and Alternative Control Strategies. In *Foodborne Diseases*; Holban, A.M., Grumezescu, A.M., Eds.; Academic Press: Cambridge, MA, USA, 2018; pp. 309–377.
14. Sereno, M.; Ziech, R.; Druziani, J.; Pereira, J.; Bersot, L. Antimicrobial Susceptibility and Biofilm Production by Salmonella sp. Strains Isolated from Frozen Poultry Carcasses. *Braz. J. Poult. Sci.* **2017**, *19*, 103–108. [CrossRef]

15. Hall-Stoodley, L.; Stoodley, P. Evolving concepts in biofilm infections. *Cell. Microbiol.* **2009**, *11*, 1034–1043. [CrossRef] [PubMed]
16. Ziech, R.E.; Perin, A.P.; Lampugnani, C.; Sereno, M.J.; Viana, C.; Soares, V.M.; Pereira, J.G.; Pinto, J.P.d.A.N.; Bersot, L.d.S. Biofilm-producing ability and tolerance to industrial sanitizers in Salmonella spp. isolated from Brazilian poultry processing plants. *LWT Food Sci. Technol.* **2016**, *68*, 85–90. [CrossRef]
17. Lamas, A.; Regal, P.; Vázquez, B.; Miranda, J.M.; Cepeda, A.; Franco, C.M. Salmonella and Campylobacter biofilm formation: A comparative assessment from farm to fork. *J. Sci. Food Agric.* **2018**, *98*, 4014–4032. [CrossRef] [PubMed]
18. Rossi, D.A.; Melo, R.T.; Mendonça, E.P.; Monteiro, G.P. Biofilms of Salmonella and Campylobacter in the poultry industry. In *Poultry Science*; IntechOpen: London, UK, 2017.
19. Turki, Y.; Ouzari, H.; Mehri, I.; Ben Aissa, R.; Hassen, A. Biofilm formation, virulence gene and multi-drug resistance in Salmonella Kentucky isolated in Tunisia. *Food Res. Int.* **2012**, *45*, 940–946. [CrossRef]
20. Bell, D.D.; Weaver, W.D. *Commercial Chicken Meat and Egg Production*, 5th ed.; Springer: New York, NY, USA, 2002; Volume 11, pp. 224–225.
21. Russell, S.M. *Controlling Salmonella in Poultry Production and Processing*; Crc Press-Taylor & Francis Group: Boca Raton, FL, USA, 2012.
22. FSIS. Modernization of Poultry Slaughter Inspection. 9 CFR Parts 381 and 500; 2014. Available online: https://www.fsis.usda.gov/wps/wcm/connect/00ffa106-f373-437a-9cf3-6417f289bfc2/2011-0012.pdf?MOD=AJPERES (accessed on 20 July 2019).
23. Rasekh, J.; Thaler, A.M.; Engeljohn, D.L.; Pihkala, N.H. Food Safety and Inspection Service policy for control of poultry contaminated by digestive tract contents: A review. *J. Appl. Poult. Res.* **2005**, *14*, 603–611. [CrossRef]
24. USDA. Related Documents for FSIS Directive 7120.1—Safe and Suitable Ingredients Used in the Production of Meat, Poultry, and Egg Products. 2019. Available online: https://www.fsis.usda.gov/wps/portal/fsis/topics/regulations/directives/7000-series/safe-suitable-ingredients-related-document (accessed on 20 July 2019).
25. Smith, J.; Corkran, S.; McKee, S.R.; Bilgili, S.F.; Singh, M. Evaluation of post-chill applications of antimicrobials against Campylobacter jejuni on poultry carcasses. *J. Appl. Poult. Res.* **2015**, *24*, 451–456. [CrossRef]
26. Duan, D.B.; Wang, H.H.; Xue, S.W.; Li, M.; Xu, X.L. Application of disinfectant sprays after chilling to reduce the initial microbial load and extend the shelf-life of chilled chicken carcasses. *Food Control* **2017**, *75*, 70–77. [CrossRef]
27. Chen, X.; Bauermeister, L.J.; Hill, G.N.; Singh, M.; Bilgili, S.F.; McKee, S.R. Efficacy of various antimicrobials on reduction of *Salmonella* and *Campylobacter* and quality attributes of ground chicken obtained from poultry parts treated in a postchill decontamination tank. *J. Food Prot.* **2014**, *77*, 1882–1888. [CrossRef]
28. Morente, E.O.; Fernandez-Fuentes, M.A.; Burgos, M.J.G.; Abriouel, H.; Pulido, R.P.; Galvez, A. Biocide tolerance in bacteria. *Int. J. Food Microbiol.* **2013**, *162*, 13–25. [CrossRef] [PubMed]
29. Gadea, R.; Fuentes, M.A.F.; Pulido, R.P.; Galvez, A.; Ortega, E. Effects of exposure to quaternary-ammonium-based biocides on antimicrobial susceptibility and tolerance to physical stresses in bacteria from organic foods. *Food Microbiol.* **2017**, *63*, 58–71. [CrossRef] [PubMed]
30. Mavri, A.; Mozina, S.S. Development of antimicrobial resistance in *Campylobacter jejuni* and *Campylobacter coli* adapted to biocides. *Int. J. Food Microbiol.* **2013**, *160*, 304–312. [CrossRef] [PubMed]
31. Capita, R.; Alvarez-Fernandez, E.; Fernandez-Buelta, E.; Manteca, J.; Alonso-Calleja, C. Decontamination treatments can increase the prevalence of resistance to antibiotics of *Escherichia coli* naturally present on poultry. *Food Microbiol.* **2013**, *34*, 112–117. [CrossRef] [PubMed]
32. Finstad, S.; O'Bryan, C.A.; Marcy, J.A.; Crandall, P.G.; Ricke, S.C. Salmonella and broiler processing in the United States: Relationship to foodborne salmonellosis. *Food Res. Int.* **2012**, *45*, 789–794. [CrossRef]
33. Chylkova, T.; Cadena, M.; Ferreiro, A.; Pitesky, M. Susceptibility of *Salmonella* biofilm and planktonic bacteria to common disinfectant agents used in poultry processing. *J. Food Prot.* **2017**, *80*, 1072–1079. [CrossRef]
34. Shah, J.; Desai, P.T.; Chen, D.; Stevens, J.R.; Weimer, B.C. Preadaptation to cold stress in *Salmonella enterica* serovar Typhimurium increases survival during subsequent acid stress exposure. *Appl. Environ. Microbiol.* **2013**, *79*, 7281–7289. [CrossRef]
35. Sun, J.; Deng, Z.; Yan, A. Bacterial multidrug efflux pumps: Mechanisms, physiology and pharmacological exploitations. *Biochem. Biophys. Res. Commun.* **2014**, *453*, 254–267. [CrossRef]

36. Molina-Gonzalez, D.; Alonso-Calleja, C.; Alonso-Hernando, A.; Capita, R. Effect of sub-lethal concentrations of biocides on the susceptibility to antibiotics of multi-drug resistant *Salmonella enterica* strains. *Food Control* **2014**, *40*, 329–334. [CrossRef]

37. Alav, I.; Rahman, K.M.; Sutton, J.M. Role of bacterial efflux pumps in biofilm formation. *J. Antimicrob. Chemother.* **2018**, *73*, 2003–2020. [CrossRef]

38. Su, C.-C.; Yin, L.; Kumar, N.; Dai, L.; Radhakrishnan, A.; Bolla, J.R.; Lei, H.-T.; Chou, T.-H.; Delmar, J.A.; Rajashankar, K.R.; et al. Structures and transport dynamics of a *Campylobacter jejuni* multidrug efflux pump. *Nat. Commun.* **2017**, *8*, 171. [CrossRef] [PubMed]

39. Jiang, J.; Xiong, Y.L.L. Technologies and mechanisms for safety control of ready-to-eat muscle foods: An updated review. *Crit. Rev. Food Sci. Nutr.* **2015**, *55*, 1886–1901. [CrossRef] [PubMed]

40. Wessels, S.; Ingmer, H. Modes of action of three disinfectant active substances: A review. *Regul. Toxicol. Pharmacol.* **2013**, *67*, 456–467. [CrossRef] [PubMed]

41. Su, X.W.; D'Souza, D.H. Reduction of *Salmonella* Typhimurium and *Listeria monocytogenes* on produce by trisodium phosphate. *LWT Food Sci. Technol.* **2012**, *45*, 221–225. [CrossRef]

42. Strempel, N.; Nusser, M.; Neidig, A.; Brenner-Weiss, G.; Overhage, J. The oxidative stress agent hypochlorite stimulates c-di-GMP synthesis and biofilm formation in *Pseudomonas aeruginosa*. *Front. Microbiol.* **2017**, *8*, 15. [CrossRef] [PubMed]

43. Zhang, C.Z.; Chen, P.X.; Yang, L.; Li, W.; Chang, M.X.; Jiang, H.X. Coordinated expression of acrAB-tolC and eight other functional efflux pumps through activating ramA and marA in *Salmonella enterica* serovar Typhimurium. *Microb. Drug Resist.* **2018**, *24*, 120–125. [CrossRef] [PubMed]

44. Urdaneta, V.; Casadesus, J. Adaptation of *Salmonella enterica* to bile: Essential role of AcrAB-mediated efflux. *Environ. Microbiol.* **2018**, *20*, 1405–1418. [CrossRef]

45. Balcázar, J.L.; Subirats, J.; Borrego, C.M. The role of biofilms as environmental reservoirs of antibiotic resistance. *Front. Microbiol.* **2015**, *6*, 1216. [CrossRef]

46. Doroshenko, N.; Tseng, B.S.; Howlin, R.P.; Deacon, J.; Wharton, J.A.; Thurner, P.J.; Gilmore, B.F.; Parsek, M.R.; Stoodley, P. Extracellular DNA Impedes the Transport of Vancomycin in Staphylococcus epidermidis Biofilms Preexposed to Subinhibitory Concentrations of Vancomycin. *Antimicrob. Agents Chemother.* **2014**, *58*, 7273–7282. [CrossRef]

47. Özdemir, C.; Akçelik, N.; Akçelik, M. The Role of Extracellular DNA in Salmonella Biofilms. *Mol. Genet. Microbiol. Virol.* **2018**, *33*, 60–71. [CrossRef]

48. Jia, K.; Wang, G.Y.; Liang, L.J.; Wang, M.; Wang, H.H.; Xu, X.L. Preliminary transcriptome analysis of mature biofilm and planktonic cells of *Salmonella* Enteritidis exposure to acid stress. *Front. Microbiol.* **2017**, *8*, 1861. [CrossRef] [PubMed]

49. Álvarez-Ordóñez, A.; Prieto, M.; Bernardo, A.; Hill, C.; López, M. The Acid Tolerance Response of Salmonella spp.: An adaptive strategy to survive in stressful environments prevailing in foods and the host. *Food Res. Int.* **2012**, *45*, 482–492. [CrossRef]

50. MacKenzie, K.D.; Wang, Y.; Shivak, D.J.; Wong, C.S.; Hoffman, L.J.L.; Lam, S.; Kröger, C.; Cameron, A.D.S.; Townsend, H.G.G.; Köster, W.; et al. Bistable expression of CsgD in *Salmonella enterica* serovar Typhimurium connects virulence to persistence. *Infect. Immun.* **2015**, *83*, 2312–2326. [CrossRef] [PubMed]

51. Wideman, N.; Bailey, M.; Bilgili, S.F.; Thippareddi, H.; Wang, L.; Bratcher, C.; Sanchez-Plata, M.; Singh, M. Evaluating best practices for *Campylobacter* and *Salmonella* reduction in poultry processing plants. *Poult. Sci.* **2016**, *95*, 306–315. [CrossRef] [PubMed]

52. Biswal, B.K.; Khairallah, R.; Bibi, K.; Mazza, A.; Gehr, R.; Masson, L.; Frigon, D. Impact of UV and peracetic acid disinfection on the prevalence of virulence and antimicrobial resistance genes in uropathogenic *Escherichia coli* in wastewater effluents. *Appl. Environ. Microbiol.* **2014**, *80*, 3656–3666. [CrossRef] [PubMed]

53. Karaca, B.; Akcelik, N.; Akcelik, M. Biofilm-producing abilities of Salmonella strains isolated from Turkey. *Biologia* **2013**, *68*, 1–10. [CrossRef]

54. Freeman, D.J.; Falkiner, F.R.; Keane, C.T. New method for detecting slime production by coagulase negative staphylococci. *J. Clin. Pathol.* **1989**, *42*, 872–874. [CrossRef]

55. Kırmusaoğlu, S. *Antimicrobials, Antibiotic Resistance, Antibiofilm Strategies and Activity Methods*; IntechOpen: London, UK, 2019.

56. Wang, H.H.; Ye, K.P.; Wei, X.R.; Cao, J.X.; Xu, X.L.; Zhou, G.H. Occurrence, antimicrobial resistance and biofilm formation of Salmonella isolates from a chicken slaughter plant in China. *Food Control* **2013**, *33*, 378–384. [CrossRef]

57. González-Rivas, F.; Ripolles-Avila, C.; Fontecha-Umaña, F.; Ríos-Castillo, A.G.; Rodríguez-Jerez, J.J. Biofilms in the Spotlight: Detection, Quantification, and Removal Methods. *Compr. Rev. Food Sci. Food Saf.* **2018**, *17*, 1261–1276. [CrossRef]

58. Romanova, N.A.; Gawande, P.V.; Brovko, L.Y.; Griffiths, M.W. Rapid methods to assess sanitizing efficacy of benzalkonium chloride to Listeria monocytogenes biofilms. *J. Microbiol. Methods* **2007**, *71*, 231–237. [CrossRef]

59. Lavilla Lerma, L.; Benomar, N.; Gálvez, A.; Abriouel, H. Prevalence of bacteria resistant to antibiotics and/or biocides on meat processing plant surfaces throughout meat chain production. *Int. J. Food Microbiol.* **2013**, *161*, 97–106. [CrossRef] [PubMed]

60. Arnold, J.W.; Silvers, S. Comparison of poultry processing equipment surfaces for susceptibility to bacterial attachment and biofilm formation. *Poult. Sci.* **2000**, *79*, 1215–1221. [CrossRef] [PubMed]

61. Okabe, S.; Hirata, K.; Watanabe, Y. Dynamic changes in spatial microbial distribution in mixed-population biofilms: Experimental results and model simulation. *Water Sci. Technol.* **1995**, *32*, 67–74. [CrossRef]

62. Itram Higiene. BioFinder. 2019. Available online: https://biofilmremove.com/en/detection/ (accessed on 22 June 2019).

63. REALCO. Biofilm Detection Kit. 2019. Available online: https://www.realco.be/en/our-markets/food-beverage/biofilm-audit/biofilm-detection-kit/ (accessed on 26 June 2019).

64. Betelgeux-Christeyns. Detection. 2019. Available online: https://biofilmtest.com/detection/?lang=en (accessed on 26 June 2019).

65. bioMérieux. TEMPO. 2019. Available online: https://www.biomerieux.com/en/biomerieux-launches-tempo-eb-first-automated-test-enterobacteriaceae-enumeration-food-products (accessed on 26 June 2019).

66. Van Houdt, R.; Michiels, C. Biofilm formation and the food industry, a focus on the bacterial outer surface. *J. Appl. Microbiol.* **2010**, *109*, 1117–1131. [CrossRef] [PubMed]

67. Lelieveld, H.L.M.; Mostert, M.A.; Holah, J. *Handbook of Hygiene Control in the Food Industry*; Woodhead Publishing Ltd.: Cambridge, UK, 2005.

68. Maukonen, J.; Mättö, J.; Wirtanen, G.; Raaska, L.; Mattila-Sandholm, T.; Saarela, M. Methodologies for the characterization of microbes in industrial environments: A review. *J. Ind. Microbiol. Biotechnol.* **2003**, *30*, 327–356. [CrossRef] [PubMed]

69. Sy-Lab. BacTrac 4300 Microbiological Impedance Analyser. 2019. Available online: https://microbiology.sylab.com/products/p/show/Product/product/bactrac-4300.html (accessed on 26 June 2019).

70. Wirtanen, G.; Salo, S.; Helander, I.M.; Mattila-Sandholm, T. Microbiological methods for testing disinfectant efficiency on Pseudomonas biofilm. *Colloids Surf. B Biointerfaces* **2001**, *20*, 37–50. [CrossRef]

71. Abcam. XTT Sodium Salt, Tetrazolium Salt (ab146310). 2019. Available online: https://www.abcam.com/xtt-sodium-salt-tetrazolium-salt-ab146310.html (accessed on 26 June 2019).

72. Abcam. Resazurin Assay Kit (Cell Viability) (ab129732). 2019. Available online: https://www.abcam.com/resazurin-assay-kit-cell-viability-ab129732.html (accessed on 26 June 2019).

73. Olson, M.E.; Ceri, H.; Morck, D.W.; Buret, A.G.; Read, R.R. Biofilm bacteria: Formation and comparative susceptibility to antibiotics. *Can. J. Vet. Res.* **2002**, *66*, 86–92. [PubMed]

74. Sarjit, A.; Dykes, G.A. Antimicrobial activity of trisodium phosphate and sodium hypochlorite against Salmonella biofilms on abiotic surfaces with and without soiling with chicken juice. *Food Control* **2017**, *73*, 1016–1022. [CrossRef]

75. Korber, D.R.; Choi, A.; Wolfaardt, G.M.; Ingham, S.C.; Caldwell, D.E. Substratum topography influences susceptibility of Salmonella enteritidis biofilms to trisodium phosphate. *Appl. Environ. Microbiol.* **1997**, *63*, 3352–3358. [PubMed]

76. Wang, H.H.; Zhang, X.X.; Dong, Y.; Xu, X.L.; Zhou, G.H. Insights into the transcriptome profile of mature biofilm of Salmonella Typhimurium on stainless steels surface. *Food Res. Int.* **2015**, *77*, 378–384. [CrossRef]

77. Borges, K.A.; Furian, T.Q.; de Souza, S.N.; Menezes, R.; de Lima, D.A.; Fortes, F.B.B.; Salle, C.T.P.; Moraes, H.L.S.; Nascimento, V.P. Biofilm formation by Salmonella Enteritidis and Salmonella Typhimurium isolated from avian sources is partially related with their in vivo pathogenicity. *Microb. Pathog.* **2018**, *118*, 238–241. [CrossRef] [PubMed]

78. Lamas, A.; Regal, P.; Vázquez, B.; Miranda, J.M.; Cepeda, A.; Franco, C.M. Influence of milk, chicken residues and oxygen levels on biofilm formation on stainless steel, gene expression and small RNAs in Salmonella enterica. *Food Control* **2018**, *90*, 1–9. [CrossRef]

79. Wang, H.; Dong, Y.; Wang, G.; Xu, X.; Zhou, G. Effect of growth media on gene expression levels in Salmonella Typhimurium biofilm formed on stainless steel surface. *Food Control* **2016**, *59*, 546–552. [CrossRef]

80. Lianou, A.; Koutsoumanis, K.P. Strain variability of the biofilm-forming ability of Salmonella enterica under various environmental conditions. *Int. J. Food Microbiol.* **2012**, *160*, 171–178. [CrossRef] [PubMed]

81. Vestby, L.K.; Møretrø, T.; Langsrud, S.; Heir, E.; Nesse, L.L. Biofilm forming abilities of Salmonellaare correlated with persistence in fish meal- and feed factories. *Bmc Vet. Res.* **2009**, *5*, 20. [CrossRef]

82. Aksoy, D. DETERMINATION OF in vitro BIOFILM FORMATION ABILITIES OF FOOD BORNE Salmonella enterica ISOLATES. *Trak. Univ. J. Nat. Sci.* **2019**, *20*, 57–62. [CrossRef]

83. Anonymous. The increasing challenge of biofilms. *Int. Food Hyg.* **2008**, *18*, 11–12.

84. Ju, X.; Li, J.; Zhu, M.; Lu, Z.; Lv, F.; Zhu, X.; Bie, X. Effect of the luxS gene on biofilm formation and antibiotic resistance by Salmonella serovar Dublin. *Food Res. Int.* **2018**, *107*, 385–393. [CrossRef]

85. Ripolles-Avila, C.; Hascoet, A.S.; Guerrero-Navarro, A.E.; Rodriguez-Jerez, J.J. Establishment of incubation conditions to optimize the in vitro formation of mature Listeria monocytogenes biofilms on food-contact surfaces. *Food Control* **2018**, *92*, 240–248. [CrossRef]

86. Giaouris, E.; Heir, E.; Hébraud, M.; Chorianopoulos, N.; Langsrud, S.; Møretrø, T.; Habimana, O.; Desvaux, M.; Renier, S.; Nychas, G.-J. Attachment and biofilm formation by foodborne bacteria in meat processing environments: Causes, implications, role of bacterial interactions and control by alternative novel methods. *Meat Sci.* **2014**, *97*, 298–309. [CrossRef]

87. Mukhopadhyay, S.; Ramaswamy, R. Application of emerging technologies to control Salmonella in foods: A review. *Food Res. Int.* **2012**, *45*, 666–677. [CrossRef]

Review

Current Knowledge on *Listeria monocytogenes* Biofilms in Food-Related Environments: Incidence, Resistance to Biocides, Ecology and Biocontrol

Pedro Rodríguez-López [1], Juan José Rodríguez-Herrera [1], Daniel Vázquez-Sánchez [2] and Marta López Cabo [1,*]

[1] Department of Microbiology and Technology of Marine Products (MICROTEC), Instituto de Investigaciones Marinas (IIM-CSIC), 6, Eduardo Cabello, 36208 Vigo, Spain; pedrorodriguez@iim.csic.es (P.R.-L.); juanherrera@iim.csic.es (J.J.R.-H.)

[2] "Luiz de Queiroz" College of Agriculture (ESALQ), University of São Paulo (USP), 11, Av. Pádua Dias, 13418-900 São Paulo, Brazil; danielvazquezsanchez@gmail.com

* Correspondence: marta@iim.csic.es; Tel.: +34-986-231-930

Received: 20 April 2018; Accepted: 1 June 2018; Published: 5 June 2018

Abstract: Although many efforts have been made to control *Listeria monocytogenes* in the food industry, growing pervasiveness amongst the population over the last decades has made this bacterium considered to be one of the most hazardous foodborne pathogens. Its outstanding biocide tolerance capacity and ability to promiscuously associate with other bacterial species forming multispecies communities have permitted this microorganism to survive and persist within the industrial environment. This review is designed to give the reader an overall picture of the current state-of-the-art in *L. monocytogenes* sessile communities in terms of food safety and legislation, ecological aspects and biocontrol strategies.

Keywords: bacteriocins; biocides; biofilm; food industry; food safety; *Listeria monocytogenes*; resistance

1. *Listeria monocytogenes*, a Food Safety Concern

Listeria monocytogenes is a ubiquitous pathogen that can stem from a febrile gastroenteritis to a severe invasive illness (listeriosis), leading to septicaemia, encephalitis, endocarditis, meningitis, abortions and stillbirths, among others syndromes [1,2]. The incidence of listeriosis is low amongst the general population, with 0.46 and 0.24 cases per 100,000 population in 2015 in the European Union and the United States respectively [3,4]. However, *L. monocytogenes* was responsible for many foodborne outbreaks with high hospitalisation and mortality rates worldwide, especially affecting pregnant women, the elderly and individuals with compromised immune systems. In particular, *L. monocytogenes* caused more foodborne outbreaks between 2005 and 2015 in the EU (83) than in the US (47), resulting in 757 and 491 cases, respectively [5–13]. In contrast, a higher number of cases required hospitalisation in the US (428) than in the EU (332), leading to more deceases (82 and 61 deaths respectively) and a higher mortality rate (24 and 16% of deceases related to foodborne outbreaks in the US and in the EU, respectively).

In spite of modifications to legal regulations, ready-to-eat (RTE) meats and dairy products are still the predominant vehicles involved in the main listeriosis outbreaks which have occurred since 2008, as well as other "low risk" products such as fruit and vegetables (Table 1). In addition to this, no consensus has been achieved among international food authorities in regards to the microbial criteria for *L. monocytogenes* [14]. As a matter of example, The United States Department of Agriculture (USDA) and the Food and Drug Administration (FDA) implemented a "zero tolerance" policy for *L. monocytogenes* contamination of RTE food products [15,16]. In contrast, the European Commission

Regulation No. 2073/2005 (amended by EC No. 1441/2007) permits levels of *L. monocytogenes* up to 100 CFU/g in RTE foods placed on the market during their shelf-life, except in those intended for infants or for special medical purposes, in which must be absent in 25 g of product [17,18]. Canada, Australia and New Zealand also permit levels of *L. monocytogenes* lower than 100 CFU/g for RTE foods in which the growth of this pathogen is limited throughout the stated shelf-life, but it must be absent in 25 g of those which can support its growth [16,19,20]. According to the Chinese Centre for Food Safety (CFS) levels of *L. monocytogenes* of 10–100 CFU/g are allowed in RTE commercialised in China, except in those refrigerated (not frozen) in which it must be absent in 25 g of product [21,22]. In Brazil, the use of *L. monocytogenes* as microbial criteria is limited to cheese, in which it must be absent in 25 g of product [23]. Curiously, many food companies follow the national regulations for products commercialised in their own country, but not foreign regulations for products that they export, leading to products with different standards of quality and safety. These actions can involve eventual problems of cross-contamination between the processing chains and a serious risk to consumers due to this lack of universal legislation. Therefore, an international consensus in microbial criteria for foodstuffs must be reached.

Table 1. Main outbreaks of foodborne listeriosis since 2008.

Year	Country	Food Product	Cases	Hospitalisations	Deaths	Ref.
2008	Canada	Delicatessen meat	57	47	24	[24]
2009–2010	Austria, Germany and Czech Republic	"Quargel" cheese	34	34	8	[25]
2009–2012	Portugal	Fresh cheeses	30	30	11	[26]
2010	Texas (US)	Diced celery	10	10	5	[27]
2011–2012	28 US states	Cantaloupes	147	143	33	[28]
2012	14 US states	Brand ricotta salata cheese	22	20	4	[29]
2012	Spain	Latin-style fresh cheese	2	2	2	[30]
2013–2014	Switzerland	RTE salad	32	32	4	[31]
2013–2014	Denmark	RTE meat products	41	41	17	[32]
2014–2015	12 US states	Caramel apples	35	34	7	[33]
2015	10 US states	Soft cheeses	30	28	3	[34]
2016	9 US states	Packaged salads	19	19	1	[3]
2016	4 US states	Frozen vegetables	9	9	3	[13]

RTE: ready-to-eat.

In the food industry, *L. monocytogenes* can persist for months or even years on floors and equipment and in drains of food-processing facilities [35,36]. This is mainly due to its ability to survive under food-related conditions that are stressful for other bacteria, such as refrigerated temperatures, desiccation, heat and high salt content [37–40], and its ability to form biofilms [41,42]. The application of ineffective cleaning and disinfection procedures in food-processing environments, particularly in locations of difficult access, also increases the risk of establishment and growth of *L. monocytogenes* and, thus, generate continuous food product contamination [43,44]. The identification of particular niches in a food-processing facility, the validation of the efficacy of sanitation procedures applied and the continuous monitoring of the presence and reestablishment in food-processing environments are therefore required to improve the control of *L. monocytogenes*.

Livestock and produce farms are considered potential primary sources for the introduction of human pathogenic *L. monocytogenes* into the food chain and food-processing plants. In fact, *L. monocytogenes* was detected in cattle, silage, animal feeds, manure and growing grass, among others [45–48]. Nevertheless, soil, water and vegetation of natural and urban environments can also serve as reservoirs of *L. monocytogenes* [49–51].

L. monocytogenes involved in most human listeriosis cases has been isolated from RTE foods post-processed in retail facilities [52,53]. The application of inadequate post-processing procedures such as improper manipulation (e.g., bacterial transfer from operator's hands and gloves, cutting boards or scales among others) or the use of contaminated slicing machines were the main cause of contamination in RTE foods [53–56]. In addition, *L. monocytogenes* is also found on non-food contact surfaces such as floors, drains, sinks, and walk-in cooler shelves of retail facilities [57,58]. *L. monocytogenes* can also proliferate due to temperature fluctuations in coolers during distribution and commercialisation of food products [59]. Moreover, this pathogen is detected in domestic environments [60–62] and public restaurants [63–65]. Several listeriosis outbreaks are also associated with foods purchased from or provided in hospitals and health care centres [66–68]. A limited knowledge of food safety, as well as an inappropriate attitude and hygiene of food handlers can directly affect the quality of the product [69,70]. Therefore, guidelines for prevention of *L. monocytogenes* contamination and persistence should be widely spread.

2. Efficacy of Food Industry Sanitisers against *L. monocytogenes*

According to published data, in Europe, around five trillion euros are invested annually for the implementation of hygienisation systems in food-related industrial environments. Nevertheless, the levels of bacterial contamination in processed food products is still a major issue of concern, with the increasing incidence of *L. monocytogenes* being remarkable if we take into account the notified cases of listeriosis [71]. The current tolerance to disinfectants in *L. monocytogenes* has been a topic of concern in the context of the food industry and public health regarding foodborne pathogens. The presence of high bacterial concentrations and the interference with organic matter due to insufficient cleaning prior to disinfection diminishes the activity and thus the efficacy of disinfectants commonly used in industrial premises [72]. This does not necessary mean that the quantity used is lower, but that the effective concentration of the antimicrobial is less than expected, compared to the initial amount deployed. However, anthropologic factors such as failure in dosage or inadequate rinsing are also responsible for the generation of tolerances due to the formation of areas in which sub-lethal concentrations of the disinfectant are present [73]. Additionally, it has also been stated that tolerance to certain disinfectants may contribute to the persistence of *L. monocytogenes* in the food industry [74].

In this section, the behaviour and further tolerance mechanisms to quaternary ammonium, chlorine and acid compounds in *L. monocytogenes*, are reviewed.

2.1. Quaternary Ammonium Compounds

Among biocides, quaternary ammonium compounds (QACs) are undoubtedly, one of the most commonly used disinfectants in the food industry efficient against bacteria, algae, fungi, spores, viruses and mycobacteria even at low concentrations [75]. QACs are active in the membrane of bacteria, casing disruption in the phospholipid bilayer and subsequent cellular content leakage causing eventual bacterial death [75]. They are stable, surface-active agents presenting low toxicity and little affected by organic matter, which make them very adequate for food industry purposes.

The described mechanisms underneath tolerances to QACs are diverse and are strongly influenced by the environment and the genetic background of each particular strain [76]. Considering the latter, a study carried out by Liu et al. [77] demonstrated how the presence of antimicrobials' sublethal concentrations can increase the possibility of oxidative stress of the cell due to an increasing concentration of free radicals in the cytoplasm. As a result, this can promote the activation of various genetic cascades like the apparition of de novo mutations due to the triggering of the SOS-response [78]. The overuse (or misuse) of QACs, may enhance the selection of new genetic elements that can be horizontally transferred [78,79]. This fact poses an additional element for the development of new forms of tolerances in *L. monocytogenes*, thus dwindling the number of options for treatment in industrial contexts that could finally enhance the biofilm formation to this pathogen [74].

Active efflux pumps are considered so far, the main mechanism for *L. monocytogenes* tolerance to QACs. This was early described by Aase et al. [80], demonstrating an extrusion of ethidium bromide outside the cell in BAC resistant and BAC adapted strains, which not only indicated the presence of an efflux pump but also that this mechanism is intrinsic to *L. monocytogenes* and can be activated by a sublethal exposure to BAC. Subsequent studies demonstrated that these efflux pumps are chromosomally encoded and that the exposure to QACs leads to an overexpression of them [76]. Despite the general agreement on this major strategy for QAC tolerance, there is still some controversy about the origin of the genetic determinants coding for efflux pumps. As a matter of example, Dutta et al. [79] demonstrated that in BAC-tolerant *L. monocytogenes* from various sources, the *bcrABC* cassette was present in 98.6% of isolates. Contrarily, Ebner et al. [81] identified the *qacH* as the main genetic determinant in BAC resistant isolates from different food matrices, and the lack of correlation between this genotype, the isolation source, the biofilm formation capability and the serotype. More recently, a new efflux pump, *emrE*, has been described in *L. monocytogenes* conferring cross-resistance to BAC and other antimicrobials [82].

Genetic mobile elements also play an important role in the dissemination of resistance genes among *L. monocytogenes*. Among, *bcrABC*-carrying isolates, it has been proposed that the transmission and subsequent integration into the chromosome, together with other resistance genes, has been mediated via transposon-containing plasmids [79]. In addition to this, Müller et al. [83] have described in *L. monocytogenes* the structure of Tn*6188*, harbouring the *qacH* gene. Ulterior investigation regarding this mobile element, has demonstrated that cells expressing *qacH*-encoded efflux pumps, showed increased MICs to BAC and other QACs, and also a decreased susceptibility to ethidium bromide [84].

Moreover, in *L. monocytogenes*, biofilm formation itself is a cause of increased tolerance to QACs due to the alterations in the membrane fluidity of the cell [85]. These alterations are mainly because of a decrease in the proportion of iso-C15 and anteiso-C15 branched-chain fatty acids (BCFA) together with an significant increase in the quantity of saturated fatty acids (SFA) [86]. Consequently, the membrane hydrophobicity is increased, thus promoting further adherence to surfaces [87]. Similar modifications have been described in cells exposed to sublethal concentrations of BAC [73] or to cold stress [87].

2.2. Chlorine-Based Compounds

Chlorines are cheap and straightforwardly used antimicrobials active against bacteria, fungi and algae. Different chlorine-based compounds such as sodium hypochlorite, chlorine dioxide gas or aqueous chlorine dioxide have been proven to be active against *L. monocytogenes* [88].

Due to their fast-oxidising nature, they interact with cellular membranes or penetrate directly into the cell forming N-chloro groups that react with the cellular metabolism due to the interference with key enzymes [89]. With this regard, Valderrama et al. [90] found a *L. monocytogenes* reduction of about 4 log CFU/mL in brine chilling solutions treated with 3 mg/mL chlorine dioxide with just 90 s contact time. Nevertheless, in *L. monocytogenes*, proper chlorine efficacy seems to be age-dependent since the thickness of the cell wall in young cultures is higher, thus protecting the cells from these sanitisers. In this line, El-Kest and Marth [91] demonstrated that a solution of 1 mg/mL of free chlorine during 10 min sufficed to reduce 4.27 orders of magnitude in 48-h-old *L. monocytogenes* Scott A cultures, whereas in 24-h-old cultures the reduction was only of 2.88 orders of magnitude.

Tolerance development against chlorine-based sanitisers has been described so far to be unlikely in *L. monocytogenes* cell suspensions [92,93]. However, Lundén et al. [94] showed that continuous transfers culture in increasing concentrations of sodium hypochlorite can promote the increase in MIC values of this disinfectant. Additionally, decreased activity of chlorine-based sanitisations have been observed not because of intrinsic factors but to interactions with external elements such as organic matter [91,92,95] or divalent cations [90].

In *L. monocytogenes* biofilms, the efficacy of chlorine solutions has been proven to greatly depend on the biofilm substrate. Hence, Bremer et al. [96] observed a significant higher proportion of eliminated cells of *L. monocytogenes* when grown on stainless steel coupons compared to those grown on polyvinyl chloride surfaces. These results were also in concordance with those obtained by Pan et al. [97] demonstrating higher tolerance to chlorine treatments in those biofilms grown in Teflon compared to those on stainless steel. Moreover, it has been observed that the adaptation of planktonic cells and subsequent sessile growth on stainless steel makes biofilms to be more tolerant to chlorine, independently of the subtype, cellular density of the biofilm and its morphology [98]. Additionally, some authors also described a cross-resistance in favour of tolerance to chlorine in *L. monocytogenes* biofilms previously treated with peroxide-based products, thus indicating that the mechanisms responsible for oxidising agents' tolerance may have a common nature in *L. monocytogenes* [97].

The effects of chlorination in *L. monocytogenes* multispecies biofilms have also been investigated. Norwood et al. [99] showed that this pathogen is able to endure concentrations higher than 1000 ppm of free chlorine in a continuous co-culture with *Staphylococcus xylosus* and *Pseudomonas fragi* on stainless steel. In contrast, other authors have found that in co-culture with *Flavobacterium* spp., a slightly acidic solution containing 400 ppm of free chlorine is enough to reduce the load both bacteria up to undetectable levels [96].

2.3. Acid Compounds

Acids are strong oxidisers able to interfere with cellular phospholipid bilayers and cytosolic material causing irreversible damage (e.g., disruption of proton motive force) and subsequent death to cells [100,101]. However, *L. monocytogenes* is able to adapt to low pH environments generated by natural processes (e.g., lactic fermentation) or artificially induced (e.g., acidification of water for cleaning systems) by means of different mechanisms. This not only allows this pathogen to survive in the environment, but could also increases its virulence since it further helps the bacterium to survive into the gastrointestinal tract and macrophage phagosome [102].

In spite of the fact that acid adaptation is a transient state in *L. monocytogenes* [103], it enhances the survival of this pathogen in the food industry, while also providing the bacterium with higher protection against other environmental insults [103]. Following this line, Phan-Thanh et al. [104] demonstrated that pre-exposure to mild acidic conditions (pH 5.5, 2 h) increased its endurance against lethal acidic, temperature (52 °C), salinity (25–30% NaCl) and alcoholic (15%) shocks. These effects are even more evident when the acid adaptation takes places gradually [102,105]. Additionally, it has been demonstrated that sublethal acid adaptation deeply alters the intracellular protein pattern expression, being more evident as the pH decreases [104,106], and that this differential pattern is strain-dependent [104].

There are different ways described in the literature in which *L. monocytogenes* can adapt to acidic conditions, all of them focused on the maintenance of the intracellular homeostasis. Among them, the glutamate decarboxylase (GAD) system is considered one of the major mechanisms [107]. This involves the GAD enzyme, which promotes the irreversible conversion of cytosolic glutamate to a neutral compound, the γ-aminobutyrate (GABA), by irreversible decarboxylation of the first [103]. The synthesis of GABA has a dual protective role: firstly, it consumes an intracellular proton during the process, with the subsequent increase of the pH inside of the cell [103]. Additionally, the extrusion of GABA outside the cell via a glutamate:GABA antiporter, contributes to the slight neutralisation of the pH outside the cell and the restarting of the metabolic pathway [102]. In food systems, it has been demonstrated that in glutamate-rich products, the survival rate of *L. monocytogenes* is significantly improved [107]. In addition to glutamate:GABA antiporter, other proton pumps such as F_0F_1-ATPase have also been proposed as active mechanisms to maintain homeostasis in acidified environments [108].

Similarly with exposure to QACs, acidic conditions modify the composition of the cytoplasmic membrane, altering the iso- and anteiso-BCFAs ratio. Giotis et al. [109] tested the response of *L. monocytogenes* 10403S to mild acid conditions, demonstrating that the total anteiso/iso ratio increased as the culture pH decreased. This was further demonstrated by Zhang et al. [110] in *L. monocytogenes* cultured in presence of various organic acids, concluding not only that the relative proportions of BCFAs were significantly altered but also that the mechanism underneath this adaptation was shared.

In biofilms, the acid-tolerance in *L. monocytogenes* seems to be strain dependent. In this line, Ibusquiza et al. [111] showed that the resistance threshold to peracetic acid between three different strains depended not only on the strain, but also the age of the biofilm and the substrate where the biofilms were grown on. These results were in accordance with those obtained by Lee et al. [112,113]. Furthermore, in addition to its overall resistance, biofilm formation is also affected when *L. monocytogenes* is exposed to acid compounds generally enhancing its adherence [114,115] even though there is evidence that early exposure to acidic conditions does not modify the ulterior biofilm formation capacity [116]. Additionally, accompanying strains, such as lactic acid bacteria, can exert a protective effect to *L. monocytogenes* in mixed-species biofilms, increasing its tolerance to acidic sanitisers [117].

3. Microbial Interactions and Resistance of *L. monocytogenes* Mixed-Species Biofilms

It is accepted that bacteria live in nature associated with another species forming structured multispecies biofilms [118]. Their life in communities makes unavoidable interspecies interaction and its impacts biofilm ecology.

Microbial communities can be defined as multispecies associations with complex structures that normally suppose important ecological advantages to the individual species present. In fact, it is accepted that biofilms represent a microbial phenotype with an explicit organisation level in which microorganisms are involved in intracellular interactions that can be competitive, cooperative or even neutral, depending on the microbial species involved and the environmental conditions [119].

Interspecies interactions are especially relevant in *L. monocytogenes* considering it is considered a poor biofilm former when compared to other bacterial species [120]. Several previous studies have addressed the influence of the accompanying microbiota in the number of adhered viable cells of *L. monocytogenes* in the corresponding mixed biofilm. There is a risk associated with the increased attachment of *L. monocytogenes* on food processing surfaces precolonised by other bacterial genera. In general, the number of adhered *L. monocytogenes* was increased, decreased or unaltered depending on the accompanying bacterium [121]. As an example, Norwood and Gilmour [122] demonstrated that higher *L. monocytogenes* numbers in monocultures compared with the multispecies biofilms formed after its association with *Staphylococcus xylosus* and *Pseudomonas fragi*. Rodríguez-López et al. [123] explored the association capacity of ten different accompanying strains with *L. monocytogenes* when forming dual-species biofilms. Outcomes demonstrated a deleterious effect of several accompanying strains on *L. monocytogenes* present on biofilms in 4 out of 10 different combinations checked. On the contrary, in other studies it has been showed that accompanying strains increase the level of adherence of *L. monocytogenes* in the mixed biofilm [123–126]. In summary, literature highlights that phenomena of adhesion/aggregation between different bacterial genera cannot be predicted since different environmental conditions can be encountered within the different niches.

Generally, previously reported studies consider that the amount of adhered viable cells in biofilms is directly related with its resistance to antimicrobials [127,128]. Nevertheless, in *L. monocytogenes*, viable biomass present in the biofilm does not give any certain indication about the difficulty of this pathogen to be eliminated from a given contamination site. In fact, a study carried out by Midelet et al. [129] demonstrated that interaction of *L. monocytogenes* with *Kocuria varians* results in higher density of the first but made its detachment easier.

The specific location of *L. monocytogenes* in the mixed microbial communities seems to be crucial when thinking on the effective elimination of the cells from contaminated surfaces or foods. Sasahara and Zottola [130] described initially interactions between *Pseudomonas* sp. and *L. monocytogenes* in biofilms and claimed on the need of a primary coloniser such as *Pseudomonas* sp. for *L. monocytogenes* to attach. Curiously, subsequent confocal microscopic studies highlight that *L. monocytogenes* locates at the bottom layers of the dual biofilms with *Pseudomonas fluorescens*. Moreover, the authors argue that *L. monocytogenes* cells have to make their way towards the biofilm bottom across the matrix [131].

Specific interspecies interactions existing in nature inside the biofilm are difficult to understand because it is impossible to empirically reproduce the strategies adopted by each species of the bacterial community to finally enhance the fitness of the biofilm consortium [119]. In spite of this, several advances have been achieved, mainly referred to the role of the accompanying microbiota.

As part of biofilm fitness, resident microbiota could protect *L. monocytogenes* to external stimuli such as food processing and/or disinfection conditions. This has been a matter of concern for biofilm researchers in the last decades. However, due to the complexity associated with the experimental work within biofilms, most of the published articles had been carried out with dual-species biofilms, which can be considered an excessive simplification of the realistic situation. *Lactobacillus plantarum* protected *L. monocytogenes* from the action of BAC and peracetic acid (PAA) [117]. Similarly, Saá Ibusquiza et al. [132] also showed that the presence of *Pseudomonas putida* increased the resistance of several strains of *L. monocytogenes* to BAC and PAA.

On the other hand, a recent study carried out by Papaioannou et al. [133] demonstrated, using a more realistic approach, that *L. monocytogenes* adhesion to stainless steel decreased (<10^2 CFU/cm^2) due to co-culture with indigenous microbiota commonly found in fish industry such as *Pseudomonas* spp., Enterobacteria or sulfide-producing bacteria. Furthermore, they postulated that this adhesion impairment was possibly one of the causes of an observed increased sensitivity to two common industrial disinfectants (Hypofoam and Divosan).

In spite of the difficulty associated with this type of studies, advances in microscopic and high throughput sequencing methodologies will permit to go deeper in the knowledge of species interactions in order to improve the hygienic design and the control of foodborne pathogens.

4. Biosanitation of *L. monocytogenes* Biofilms Using Lactic Acid Bacteria and Bacteriocins

The removal of microorganisms from food premises cannot be currently conceived without the use of conventional biocides. However, this practice has not been completely successful, and some issues have arisen. For instance, the emergence of resistant (or tolerant) strains [134,135] and a highly decreased effectiveness in the presence organic material [136] or in "harbourage sites" [121].

Huge efforts have been therefore conducted to search for new strategies of control of foodborne pathogens, with particular reference to *L. monocytogenes* [137–139]. This search has also included cost-effectiveness, environmentally friendly nature and low toxicity to humans as further major requirements. As a result, a number of promising alternatives have been identified, such a lactic acid bacteria (LAB), bacteriocins or bacteriophages, enzymes and surfactants (mainly as anti-adhesion and detachment agents), essential oils, electrolysed oxidising water and ozone, photocatalysts, ionising and UV radiation or ultrasonication, among others.

It is widely known that the microbiota present in food facilities can enhance or inhibit the colonisation of surfaces by *L. monocytogenes* [119,125,140]. Accordingly, Fox et al. [141] proposed to influence the microbiome in favour of antilisterial species as a strategy to reduce the presence of *L. monocytogenes*. However, the microorganisms are, in general, undesirable in food premises, since they may promote food spoilage or cause food safety problems. This is may not be the case for LAB, particularly for those having probiotic properties.

Many different LAB and several bacteriocins are known to be highly active against Gram-positive bacteria, particularly against *L. monocytogenes* [142,143]. In addition, the presence of antilisterial structural bacteriocins genes in LAB has been recently reported [144]. Accordingly, several studies have examined the potential of LAB and their bacteriocins as a tool for the control of biofilms of *L. monocytogenes* in food facilities. This last section of this review is intended to briefly outline some of the most significant results of these studies.

4.1. Preventing Biofilm Formation

Nisin is, without any doubt, the most studied bacteriocin. Moreover, among bacteriocins, only nisin has been granted a generally recognised as safe (GRAS) status by the FDA and approved for use as a food additive (additive number E234) within the European Union. Initial studies were, therefore, focused on the effectiveness of nisin and nisin-producing *Lactococcus lactis* subsp. *lactis* strains against biofilms. An early study by Bower et al. [145] already showed that nisin films adsorbed on silica surfaces inhibited the growth of *L. monocytogenes*. In addition, high nisin concentrations were found to be lethal to attached cells.

Since then, a number of studies have evaluated the potential of several bacteriocins to prevent adhesion and biofilm formation of *L. monocytogenes* on different plastic and metallic substrates, specifically nisin [146–148], enterocins [148,149] and sakacin 1 [150]. Although these studies were conducted using from pure bacteriocins to cell-free extracts, results clearly show that bacteriocins may delay but not prevent biofilm formation completely. In fact, only enterocin AS-48 (conditioned on polystyrene surfaces) was reported to be able to completely inhibit biofilm formation for at least 24 h, but longer times of study were not tested [149]. Similarly, Winkelströter et al. [150] also observed a noticeable inhibition of initial stages of biofilm formation for up to 24 h in the presence of a cell-free neutralised supernatant of *Lactobacillus sakei* 1 (containing non-purified sakacin 1), but regrowth of biofilms took place subsequently, which was attributed to a possible lack of competition for nutrients or a selection of bacteriocin-tolerant phenotypes.

The effect of *Lactococcus lactis* CNRZ 150, a nisin-producing strain, against the attachment of *L. monocytogenes* was also examined [151]. These authors underlined an additional advantage of using bacteriocin-producing LAB over bacteriocins, that is, competitive inhibition limiting nutrient supply and, accordingly, two different approaches were addressed. The first, "deferred adhesion" (later defined as exclusion mechanism), consists of testing the effect of pre-formed *L. lactis* biofilms. In the second, "simultaneous adhesion" (later, competition mechanism), the effect is tested by co-culturing *L. lactis* and *L. monocytogenes*. In both scenarios, attachment of *L. monocytogenes* and subsequent biofilm formation was effectively prevented.

Considering that it is highly likely that *L. monocytogenes* encounters resident biofilms rather than solid abiotic surfaces in food-processing environments [152], the effectiveness of LAB against the attachment of *L. monocytogenes* has been interchangeably tested in terms of exclusion or competition mechanisms in many subsequent studies. This effect has been defined as competitive exclusion. As a result, a high number of different strains has been shown to be highly effective: *Enterococcus durans* [153,154], *Enterococcus faecium* [146], *L. lactis* [153–155], *Lactobacillus plantarum* and *Enterococcus casseliflavus* [156], *Leuconostoc mesenteroides* [157], *L. sakei* [150,158], *Pediococcus acidilactici*, *Lactobacillus amylovorus* and *Lactobacillus animalis* [159], *Lactobacillus acidophilus*, *Lactobacillus casei*, *Lactobacillus paracasei* and *Lactobacillus rhamnosus* [160], and *Lactobacillus paraplantarum* [161], among others.

Generally, bacteriocin-producing strains have been found to be more effective than non-bacteriocin-producing strains against biofilm formation by *L. monocytogenes*. This was clearly found for *E. faecium* [146], *L. mesenteroides* [157] and *L. sakei* [150]. However, the effectiveness of LAB can be also due to other antimicrobial metabolites, such as lactic acid and other organic acids which also decrease pH, as well as biosurfactants that can additionally prevent adhesion [155,162].

Additionally, competition for adhesion sites and nutrients was also shown to inhibit biofilm formation [163,164]. Interestingly, a study performed by Habimana et al. [164] showed by confocal laser-scanning microscopy of dual-species biofilms formed by co-culture with *L. lactis* that *L. monocytogenes* cells were located in the bottom layers of biofilms, entirely covered by *L. lactis*. In addition, modelling revealed that *L. monocytogenes* would be, in their own words, smothered by competitors and forced into a survival lifestyle, rather than into proliferation or colonisation processes. This inhibition would mainly occur during the initial phases of biofilm formation, essentially due to longer generation time and latency of *L. monocytogenes*. A similar effect had been already detected by Leriche et al. [151], who found that *L. monocytogenes* became permanent resident in dual-species biofilms when the inoculum size was high (10^8 CFU/mL), even though high densities of *L. lactis* were able to outcompete and prevent *L. monocytogenes* to resume growth on surfaces. In keeping with these studies, it is worth to indicate that *L. monocytogenes* was found to be more resistant to disinfection in dual-species biofilms with *L. plantarum* than in single-species biofilms, particularly when outnumbered by *L. plantarum*, which seems to indicate a protective effect of the latter [117].

4.2. Removal of Biofilms

In line with these above mentioned studies, different comparative studies have shown that the effectiveness of LAB against pre-formed biofilms of *L. monocytogenes* (approach known as displacement mechanism) is significantly lower than on adhesion and biofilm formation by competitive exclusion [158–160]. That is, acting early would seem to be most appropriate to prevent biofilm formation. A similar conclusion can be drawn by comparing results from two different studies conducted by Zhao et al. [153,154], despite the effectiveness of displacement being considerably increased by extending treatments with LAB for up to 3 weeks.

Similarly, nisin was found to act rather slowly and, more importantly, with a limited effectiveness against pre-formed biofilms of *L. monocytogenes* [111,155]. This was attributed to a reduced ability to diffuse into the matrix and reach cells. Subsequent studies have confirmed that nisin does not seem to be practical as a surface sanitiser on its own [165,166]. Biofilms were also highly resistant to treatments with enterocin [149] and a semi-purified curvacin extract of *L. sakei* [158]. Concentrated cell-free supernatants from several bacteriocin-producing LAB did not have strong effects on pre-formed biofilms of *L. monocytogenes* either [167]. On the contrary, an important effect on biofilms was recently claimed for both nisin and enterocin [148], but rather high cell densities could be still clearly observed by scanning electron microscopy following treatments.

4.3. Combined Treatments

L. monocytogenes can develop tolerance and even resistance to bacteriocins if exposed to sub-inhibitory concentrations [168,169], and this decreases substantially the efficacy of treatments. Thus, Bower et al. [145] had already shown that coating surfaces with nisin did not inhibit the adhesion of nisin-resistant *L. lactis*. Combining LAB or bacteriocins with other antimicrobial factors may provide a greater effect, something that has been widely known for a long time. Thus, Leriche et al. [151] had already suggested the use of hurdle technology-like approaches to overcome bacteriocin resistance. Some studies have tested this strategy on biofilms of *L. monocytogenes*.

Of note, the treatment of floor drains of food-processing facilities with one strain of *L. lactis* subsp. *lactis* and other of *E. durans* greatly reduced the contamination with *L. monocytogenes* [154,170]. This combination should reduce the likelihood that *L. monocytogenes* developed tolerance to nisin too. Thus, most drains were found to remain free of detectable *L. monocytogenes* for several weeks after completing treatments.

Remarkably, bioencapsulation of thermally-treated fermentates of two strains of *Carnobacterium maltaromaticum* and one of *Enterococcus mundtii*, plus a relatively high nisin concentration, in an alginate matrix supported by a mesh-type fabric was highly effective against biofilms of *L. monocytogenes* [171]. Bioencapsulation allows bacteriocins to be slowly released, which seems to be more effective than large doses [172], as long as the emergence of resistance does not occur. This biocontroller eliminated *L monocytogenes* from biofilms formed in floor gutters in a fish processing plant after only 48 h of contact time, but was rather ineffective against biofilms formed on plastic surfaces (i.e., Teflon and rubber), where they were thinner and the attachment was stronger than on stainless steel. Importantly, conventional biocides did not reduce the effectiveness of the biocontroller. They were thus used jointly to achieve maximum effectiveness.

As an alternative to combine different LAB, some researchers have proposed to combine bacteriocins with different modes of action. This involved merging nisin—a class I bacteriocin—with enterocin—belonging to class IIb—, a bacteriocin produced by enterococci, was highly active against biofilms of nisin-resistant *L. monocytogenes*. Four-fold less of both bacteriocins were required and, importantly, no cross-resistance was detected [148]. On the contrary, cross-resistance for nisin and class IIa bacteriocins has been detected [173,174]. Nonetheless, previous studies demonstrated that some enterocins can present cytotoxicity upon epithelial cells [175], hence, whether they can be safe for use in food environments still remains to be clarified.

Combining bacteriocins with conventional biocides also seems an attractive strategy to reduce the likelihood of colonisation by resistant variants. Thus, the combination of enterocin AS-48 with different commercial sanitisers (quaternary ammonium compounds, bis-phenols or guanides) was found to be much more effective than any single treatment, but this effect was not observed with oxidising agents [149]. This approach would also allow conventional biocides to be used in lower amounts while increasing efficacy, which is highly important to reduce toxicity to humans and in the environment.

4.4. Final Considerations

A controlled application of LAB seems a very promising approach to prevent or even remove *L. monocytogenes* from food facilities basically as a result of a high competitive potential for adhesion sites and nutrients, and the production of some growth-inhibiting compounds, majorly bacteriocins. Moreover, the ability of LAB to spread and colonise surfaces can make them highly suitable as an alternative treatment for difficult-to-reach locations, where *L. monocytogenes* is not easily removed by routine cleaning and disinfection.

LAB have been safely used by humans for centuries in food production and preservation. However, they have no legal status for use as biosanitisers in the food industry. Accordingly, some issues have arisen concerning the use of some LAB. Prerequisites for a safe use need to be clearly defined.

Ideally, bacteriocin-producing LAB with no cross-resistance should be strategically combined to increase efficacy and prevent the emergence of bacteriocin-resistant phenotypes. In this sense, studies intended to validate different combinations of LAB should be encouraged. However, LAB generally join pre-existing polymicrobial biofilms in food processing environments rather than forming new structures. Consequently, this coexisting microbiota as well as temperature (which fluctuates rather significantly), the surface or soiling, among other factors, can eventually affect attachment, growth and bacteriocin production of each LAB, and therefore the effectiveness of treatments. Treatments should be therefore optimised individually. Unfortunately, only a small number of studies have addressed in situ testing [154,170,171], which makes it highly likely that applications are far from being straightforward. The design of strategies for in situ application of LAB in the food industry is thus needed.

5. Conclusions and Future Perspectives

There is no doubt that the recalcitrance of *L. monocytogenes* in foodstuffs is greatly influenced by the ubiquitous presence of its biofilm among food contact and non-food contact surfaces within food-processing premises [121]. Despite the great advances that have been made over the last few decades in the field of food safety, several outbreaks (Table 1) with high rates of morbidity and mortality, especially within the so-called YOPI (young, old, pregnant, immunosupressed) group, are still reported [71,176].

Understanding the various genetic and physiological underlying mechanisms leading to antimicrobial resistance as well as the influence on *L. monocytogenes* of pre-existing resident/transient microbiota and vice versa, are nowadays considered as key factors to developing fast, efficient, safe and cost-effective treatments in order to improve the environmental control of this foodborne pathogen.

Additionally to biocontrol as presented in this review, there is a significant amount of ongoing investigation developed by several groups focused on the design of ad hoc antibiofilm strategies such as enzymes [177], bacteriophages [178] or combined strategies [179]. Nevertheless, the rapid adaptation undergone by the different members of sessile communities makes us always being one step behind. Hence, the development of preventive rather than disinfecting strategies based on case-by-case approaches appears as wide field of research to go in-depth to eventually ensure the quality and safety of foodstuffs consumed in the society.

Author Contributions: Conceptualisation of the reivew: M.L.C. Writing (original manuscript): P.R.-L., J.J.R.-H., D.V.-S. and M.L.C. Formatting, reviewing and editing: P.R.-L.

Acknowledgments: Authors P.R.-L., J.J.R.-H. and M.L.C. acknowledge the Spanish Ministerio de Economía, Industria y Competitividad (MINEICO) for its financial support (Project: SOLISTA; AGL2016_78549). Author D.V.-S. was financially supported by a research grant of the São Paulo Research Foundation (FAPESP, 2014/20590-0).

Conflicts of Interest: The authors declare no conflict of interest.

References

1. Charlier, C.; Perrodeau, É.; Leclercq, A.; Cazenave, B.; Pilmis, B.; Henry, B.; Lopes, A.; Maury, M.M.; Moura, A.; Goffinet, F.; et al. Clinical features and prognostic factors of listeriosis: The MONALISA national prospective cohort study. *Lancet Infect. Dis.* **2017**, *17*, 510–519. [CrossRef]
2. Vázquez-Boland, J.A.; Kuhn, M.; Berche, P.; Chakraborty, T.; Domi, G.; González-zorn, B.; Wehland, J. Listeria Pathogenesis and Molecular Virulence Determinants Listeria Pathogenesis and Molecular Virulence Determinants. *Clin. Microbiol. Rev.* **2001**, *14*, 584–640. [CrossRef] [PubMed]
3. CDC. Foodborne Diseases Active Surveillance Network (FoodNet). Available online: https://www.cdc.gov/foodnet/reports/data/infections.html (accessed on 12 January 2018).
4. EFSA. The Euroean Union summary report on trends and sources of zoonsoes, zoonotic agents and food-borne outbreaks in 2015. *EFSA J.* **2016**, *14*, 4634. [CrossRef]
5. EFSA. The Community summary report on trends and sources of zoonoses, zoonotic agents, antimicrobial resistance and foodborne outbreaks in the European Union in 2005. *EFSA J.* **2006**, *94*. [CrossRef]
6. EFSA. The Community Summary Report on Trends and Sources of Zoonoses, Zoonotic Agents, Antimicrobial Resistance and Foodborne Outbreaks in the European Union in 2006. *EFSA J.* **2007**, *130*. [CrossRef]
7. EFSA. The Community summary report on trends and sources of zoonoses and zoonotic agents in the European Union in 2007. *EFSA J.* **2009**, *223*. [CrossRef]
8. EFSA. The European Union summary report on trends and sources of zoonoses, zoonotic agents and food-borne outbreaks in 2009. *EFSA J.* **2011**, *9*. [CrossRef]
9. EFSA. Trends and Sources of Zoonoses and Zoonotic Agents and Food-borne Outbreaks in 2011. *EFSA J.* **2013**, *11*, 3129. [CrossRef]
10. EFSA. The European Union Summary Report on Trends and Sources of Zoonoses, Zoonotic Agents and Food-borne Outbreaks in 2012. *EFSA J.* **2014**, *12*. [CrossRef]
11. EFSA. The European Union summary report on trends and sources of zoonoses, zoonotic agents and food-borne outbreaks in 2013. *EFSA J.* **2015**, *13*, 3991. [CrossRef]
12. EFSA. The European Union summary report on trends and sources of zoonoses, zoonotic agents and food-borne outbreaks in 2014. *EFSA J.* **2015**, *13*, 4329. [CrossRef]
13. CDC. National Outbreak Reporting System (NORS). Available online: https://www.cdc.gov/nors/data/dashboard/faq-using-dashboard.html (accessed on 20 January 2018).
14. Hoagland, L.; Ximenes, E.; Ku, S.; Ladisch, M. Foodborne pathogens in horticultural production systems: Ecology and mitigation. *Sci. Hortic.* **2018**, *236*, 192–206. [CrossRef]
15. USDA-FSIS. *Verification Activities for the Listeria monocytogenes (Lm) Regulation and the Ready-To-Eat (RTE) Sampling Program*; United States Department of Agriculture, Food Safety and Inspection Service: Washington, DC, USA, 2014; p. 48.
16. FDA. Guidelines for the Microbiological Examination of Ready-To-Eat Foods. Available online: https://www.foodstandards.gov.au/code/microbiollimits/documents/GuidelinesforMicroexam.pdf (accessed on 3 May 2018).
17. European Commission. Commission Regulation (EC) No 2073/2005 of 15th November 2005 on microbiological criteria for foodstuffs. *Off. J. Eur. Union* **2005**, *L338*, 1–26.
18. European Commission. Commission Regulation (EC) No 1441/2007 of 5 December 2007 amending Regulation (EC) No 2073/2005 on microbiological criteria for foodstuffs. *Off. J. Eur. Union* **2007**, *L322*, 12–29.
19. Policy on Listeria Monocytogenes in Ready-to-Eat (RTE) Foods. Available online: https://www.canada.ca/content/dam/hc-sc/migration/hc-sc/fn-an/alt_formats/pdf/legislation/pol/policy_listeria_monocytogenes_2011-eng.pdf (accessed on 20 January 2018).
20. Australia New Zealand Food Standards Code—Schedule 27—Microbiological Limits in Food. Available online: https://www.legislation.gov.au/Details/F2017C00323 (accessed on 20 January 2018).
21. Centre for Food Safety. *Microbiological Guidelines for Food. For Ready-To-Eat Food in General and Specific Food Items*; Food and Environmental Hygiene Department: Hong Kong, China, 2014.
22. Centre for Food Safety. Microbiological Guidelines for Ready-To-Eat Food. Available online: http://blpd.dss.go.th/micro/ready.pdf (accessed on 3 May 2018).

23. ANVISA. *Regulamento Técnico Sobre os Padrões Microbiológicos Para Alimentos*; Diário Oficial da República Federativa do Brasil: Brasília, Brazil, 2001; p. 48.

24. Thomas, M.K.; Vriezen, R.; Farber, J.M.; Currie, A.; Schlech, W.; Fazil, A. Economic Cost of a *Listeria monocytogenes* Outbreak in Canada, 2008. *Foodborne Pathog. Dis.* **2015**, *12*, 966–971. [CrossRef] [PubMed]

25. Schoder, D.; Stessl, B.; Szakmary-Brändle, K.; Rossmanith, P.; Wagner, M. Population diversity of *Listeria monocytogenes* in quargel (acid curd cheese) lots recalled during the multinational listeriosis outbreak 2009/2010. *Food Microbiol.* **2014**, *39*, 68–73. [CrossRef] [PubMed]

26. Magalhães, R.; Almeida, G.; Ferreira, V.; Santos, I.; Silva, J.; Mendes, M.M.; Pita, J.; Mariano, G.; Mancio, I.; Sousa, M.M.; et al. Cheese-related listeriosis outbreak, Portugal, March 2009 to February 2012. *Eurosurveillance* **2015**, *20*. [CrossRef]

27. Gaul, L.K.; Farag, N.H.; Shim, T.; Kingsley, M.A.; Silk, B.J.; Hyytia-Trees, E. Hospital-Acquired Listeriosis Outbreak Caused by Contaminated Diced Celery—Texas, 2010. *Clin. Infect. Dis.* **2013**, *56*, 20–26. [CrossRef] [PubMed]

28. McCollum, J.T.; Cronquist, A.B.; Silk, B.J.; Jackson, K.A.; O'Connor, K.A.; Cosgrove, S.; Gossack, J.P.; Parachini, S.S.; Jain, N.S.; Ettestad, P.; et al. Multistate Outbreak of Listeriosis Associated with Cantaloupe. *N. Engl. J. Med.* **2013**, *369*, 944–953. [CrossRef] [PubMed]

29. Heiman, K.E.; Garalde, V.B.; Gronostaj, M.; Jackson, K.A.; Beam, S.; Joseph, L.; Saupe, A.; Ricotta, E.; Waechter, H.; Wellman, A.; et al. Multistate outbreak of listeriosis caused by imported cheese and evidence of cross-contamination of other cheeses, USA, 2012. *Epidemiol. Infect.* **2016**, *144*, 2698–2708. [CrossRef] [PubMed]

30. Castro, V.; Escudero, J.M.; Rodriguez, J.L.; Muniozguren, N.; Uribarri, J.; Saez, D.; Vazquez, J. Listeriosis outbreak caused by Latin-style fresh cheese, Bizkaia, Spain, August 2012. *Eurosurveillance* **2012**, *17*, 3–5. [CrossRef]

31. Stephan, R.; Althaus, D.; Kiefer, S.; Lehner, A.; Hatz, C.; Schmutz, C.; Jost, M.; Gerber, N.; Baumgartner, A.; Hächler, H.; et al. Foodborne transmission of *Listeria monocytogenes* via ready-to-eat salad: A nationwide outbreak in Switzerland, 2013–2014. *Food Control* **2015**, *57*, 14–17. [CrossRef]

32. Jensen, A.K.; Nielsen, E.M.; Björkman, J.T.; Jensen, T.; Müller, L.; Persson, S.; Bjerager, G.; Perge, A.; Krause, T.G.; Kiil, K.; et al. Whole-genome sequencing used to investigate a nationwide outbreak of listeriosis caused by ready-to-eat delicatessen meat, Denmark, 2014. *Clin. Infect. Dis.* **2016**, *63*, 64–70. [CrossRef] [PubMed]

33. Angelo, K.M.; Conrad, A.R.; Saupe, A.; Dragoo, H.; West, N.; Sorenson, A.; Barnes, A.; Doyle, M.; Beal, J.; Jackson, K.A.; et al. Multistate outbreak of *Listeria monocytogenes* infections linked to whole apples used in commercially produced, prepackaged caramel apples: United States, 2014–2015. *Epidemiol. Infect.* **2017**, *145*, 848–856. [CrossRef] [PubMed]

34. CDC. Multistate Outbreak of Listeriosis Linked to Soft Cheeses Distributed by Karoun Dairies, Inc. Available online: http://www.cdc.gov/listeria/outbreaks/soft-cheeses-09-15/index.html (accessed on 21 January 2018).

35. Ferreira, V.; Wiedmann, M.; Teixeira, P.; Stasiewicz, M.J. *Listeria monocytogenes* persistence in food-associated environments: Epidemiology, strain characteristics, and implications for public health. *J. Food Prot.* **2014**, *77*, 150–170. [CrossRef] [PubMed]

36. Jami, M.; Ghanbari, M.; Zunabovic, M.; Domig, K.J.; Kneifel, W. *Listeria monocytogenes* in Aquatic Food Products—A Review. *Compr. Rev. Food Sci. Food Saf.* **2014**, *13*, 798–813. [CrossRef]

37. Doyle, M.E.; Mazzotta, A.S.; Wang, T.; Wiseman, D.W.; Scott, V.N. Heat resistance of *Listeria monocytogenes*. *J. Food Prot.* **2001**, *64*, 410–429. [CrossRef]

38. Gardan, R.; Cossart, P.; Labadie, J. Identification of *Listeria monocytogenes* genes involved in salt and alkaline-pH tolerance. *Appl. Environ. Microbiol.* **2003**, *69*, 3137–3143. [CrossRef] [PubMed]

39. Moorhead, S.M.; Dykes, G.A. Influence of the sigB gene on the cold stress survival and subsequent recovery of two *Listeria monocytogenes* serotypes. *Int. J. Food Microbiol.* **2004**, *91*, 63–72. [CrossRef]

40. Zoz, F.; Grandvalet, C.; Lang, E.; Iaconelli, C.; Gervais, P.; Firmesse, O.; Guyot, S.; Beney, L. *Listeria monocytogenes* ability to survive desiccation: Influence of serotype, origin, virulence, and genotype. *Int. J. Food Microbiol.* **2017**, *248*, 82–89. [CrossRef] [PubMed]

41. Møretrø, T.; Langsrud, S. *Listeria monocytogenes*: Biofilm formation and persistence in food-processing environments. *Biofilms* **2004**, *1*, 107–121. [CrossRef]

42. Doijad, S.P.; Barbuddhe, S.B.; Garg, S.; Poharkar, K.V.; Kalorey, D.R.; Kurkure, N.V.; Rawool, D.B.; Chakraborty, T. Biofilm-forming abilities of listeria monocytogenes serotypes isolated from different sources. *PLoS ONE* **2015**, *10*, e0137046. [CrossRef] [PubMed]
43. Chmielewski, R.A.N.; Frank, J.F. Biofilm Formation and Control in Food Processing Facilities. *Compr. Rev. Food Sci. Food Saf.* **2003**, *2*, 22–32. [CrossRef]
44. Martín, B.; Perich, A.; Gómez, D.; Yangüela, J.; Rodríguez, A.; Garriga, M.; Aymerich, T. Diversity and distribution of *Listeria monocytogenes* in meat processing plants. *Food Microbiol.* **2014**, *44*, 119–127. [CrossRef] [PubMed]
45. Wiedmann, M.; Bruce, J.L.; Knorr, R.; Bodis, M.; Cole, E.M.; McDowell, C.I.; McDonough, P.L.; Batt, C.A. Ribotype diversity of *Listeria monocytogenes* strains associated with outbreaks of listeriosis in ruminants. *J. Clin. Microbiol.* **1996**, *34*, 1086–1090. [PubMed]
46. Fenlon, D.R.; Wilson, J.; Donachie, W. The incidence and level of *Listeria monocytogenes* contamination of food sources at primary production and initial processing. *J. Appl. Bacteriol.* **1996**, *81*, 641–650. [CrossRef] [PubMed]
47. Nightingale, K.K.; Schukken, Y.H.; Nightingale, C.R.; Fortes, E.D.; Ho, A.J.; Her, Z.; Grohn, Y.T.; McDonough, P.L.; Wiedmann, M. Ecology and transmission of *Listeria monocytogenes* infecting ruminants and in the farm environment. *Appl. Environ. Microbiol.* **2004**, *70*, 4458–4467. [CrossRef] [PubMed]
48. Castro, H.K.; Lindström, M. Ecology and surveillance of *Listeria monocytogenes* on dairy cattle farms. *Int. J. Infect. Dis.* **2016**, *53*, 68. [CrossRef]
49. Sauders, B.D.; Overdevest, J.; Fortes, E.; Windham, K.; Schukken, Y.; Lembo, A.; Wiedmann, M. Diversity of Listeria species in urban and natural environments. *Appl. Environ. Microbiol.* **2012**, *78*, 4420–4433. [CrossRef] [PubMed]
50. Vivant, A.-L.; Garmyn, D.; Piveteau, P. *Listeria monocytogenes*, a down-to-earth pathogen. *Front. Cell. Infect. Microbiol.* **2013**, *3*, 87. [CrossRef] [PubMed]
51. Linke, K.; Rückerl, I.; Brugger, K.; Karpiskova, R.; Walland, J.; Muri-Klinger, S.; Tichy, A.; Wagner, M.; Stessl, B. Reservoirs of Listeria species in three environmental ecosystems. *Appl. Environ. Microbiol.* **2014**, *80*, 5583–5592. [CrossRef] [PubMed]
52. Lianou, A.; Sofos, J.N. A Review of the Incidence and Transmission of *Listeria monocytogenes* in Ready-to-Eat Products in Retail and Food Service Environments. *J. Food Prot.* **2007**, *70*, 2172–2198. [CrossRef] [PubMed]
53. Hoelzer, K.; Oliver, H.F.; Kohl, L.R.; Hollingsworth, J.; Wells, M.T.; Wiedmann, M. Structured Expert Elicitation about *Listeria monocytogenes* Cross-Contamination in the Environment of Retail Deli Operations in the United States. *Risk Anal.* **2012**, *32*, 1139–1156. [CrossRef] [PubMed]
54. Endrikat, S.; Gallagher, D.; Pouillot, R.; Quesenberry, H.H.; Labarre, D.; Schroeder, C.M.; Kause, J. A comparative risk assessment for *Listeria monocytogenes* in prepackaged versus retail-sliced deli meat. *J. Food Prot.* **2010**, *73*, 612–619. [CrossRef]
55. Chaitiemwong, N.; Hazeleger, W.C.; Beumer, R.R.; Zwietering, M.H. Quantification of transfer of *Listeria monocytogenes* between cooked ham and slicing machine surfaces. *Food Control* **2014**, *44*, 177–184. [CrossRef]
56. Scollon, A.M.; Wang, H.; Ryser, E.T. Transfer of *Listeria monocytogenes* during mechanical slicing of onions. *Food Control* **2016**, *65*, 160–167. [CrossRef]
57. Hoelzer, K.; Sauders, B.D.; Sanchez, M.D.; Olsen, P.T.; Pickett, M.M.; Mangione, K.J.; Rice, D.H.; Corby, J.; Stich, S.; Fortes, E.D.; et al. Prevalence, distribution, and diversity of *Listeria monocytogenes* in retail environments, focusing on small establishments and establishments with a history of failed inspections. *J. Food Prot.* **2011**, *74*, 1083–1095. [CrossRef] [PubMed]
58. Lakicevic, B.; Nastasijevic, I. *Listeria monocytogenes* in retail establishments: Contamination routes and control strategies. *Food Rev. Int.* **2017**, *33*, 247–269. [CrossRef]
59. Zeng, W.; Vorst, K.; Brown, W.; Marks, B.P.; Jeong, S.; Pérez-Rodríguez, F.; Ryser, E.T. Growth of *Escherichia coli* O157:H7 and *Listeria monocytogenes* in Packaged Fresh-Cut Romaine Mix at Fluctuating Temperatures during Commercial Transport, Retail Storage, and Display. *J. Food Prot.* **2014**, *77*, 197–206. [CrossRef] [PubMed]
60. Azevedo, I.; Regalo, M.; Mena, C.; Almeida, G.; Carneiro, L.; Teixeira, P.; Hogg, T.; Gibbs, P.A. Incidence of *Listeria* spp. in domestic refrigerators in Portugal. *Food Control* **2005**, *16*, 121–124. [CrossRef]
61. Beumer, R.R.; te Giffel, M.C.; Spoorenberg, E.; Rombouts, F.M. Listeria species in domestic environments. *Epidemiol. Infect.* **1996**, *117*, 437–442. [CrossRef] [PubMed]

62. Jackson, V.; Blair, I.S.; McDowell, D.A.; Kennedy, J.; Bolton, D.J. The incidence of significant foodborne pathogens in domestic refrigerators. *Food Control* **2007**, *18*, 346–351. [CrossRef]
63. Lin, C.-M.; Fernando, S.Y.; Wei, C. Occurrence of *Listeria monocytogenes*. *Salmonella* spp., *Escherichia coli* and *E. coli* O157:H7 in vegetable salads. *Food Control* **1996**, *7*, 135–140. [CrossRef]
64. Soriano, J.M.; Rico, H.; Moltó, J.C.; Mañes, J. Listeria species in raw and ready-to-eat foods from restaurants. *J. Food Prot.* **2001**, *64*, 551–553. [CrossRef] [PubMed]
65. Barkley, J.S.; Gosciminski, M.; Miller, A. Whole-genome sequencing detection of ongoing Listeria contamination at a restaurant, Rhode Island, USA, 2014. *Emerg. Infect. Dis.* **2016**, *22*, 1474–1476. [CrossRef] [PubMed]
66. Little, C.L.; Amar, C.F.L.; Awofisayo, A.; Grant, K.A. Hospital-acquired listeriosis associated with sandwiches in the UK: A cause for concern. *J. Hosp. Infect.* **2012**, *82*, 13–18. [CrossRef] [PubMed]
67. Cokes, C.; France, A.M.; Reddy, V.; Hanson, H.; Lee, L.; Kornstein, L.; Stavinsky, F.; Balter, S. Serving high-risk foods in a high-risk setting: Survey of hospital food service practices after an outbreak of listeriosis in a hospital. *Infect. Control Hosp. Epidemiol.* **2011**, *32*, 380–386. [CrossRef] [PubMed]
68. Martins, I.S.; Da Conceião Faria, F.C.; Miguel, M.A.L.; De Sá Colao Dias, M.P.; Cardoso, F.L.L.; De Gouveia Magalhães, A.C.; Mascarenhas, L.A.; Nouér, S.A.; Barbosa, A.V.; Vallim, D.C.; et al. A cluster of *Listeria monocytogenes* infections in hospitalized adults. *Am. J. Infect. Control* **2010**, *38*, e31–e36. [CrossRef] [PubMed]
69. Lee, H.K.; Abdul Halim, H.; Thong, K.L.; Chai, L.C. Assessment of food safety knowledge, attitude, self-reported practices, and microbiological hand hygiene of food handlers. *Int. J. Environ. Res. Public Health* **2017**, *14*, 55. [CrossRef]
70. Akabanda, F.; Hlortsi, E.H.; Owusu-Kwarteng, J. Food safety knowledge, attitudes and practices of institutional food-handlers in Ghana. *BMC Public Health* **2017**, *17*, 40. [CrossRef] [PubMed]
71. EFSA; ECDC. The European Union summary report on trends and sources of zoonoses, zoonotic agents and food-borne outbreaks in 2016. *EFSA J.* **2017**, *15*, 5077. [CrossRef]
72. Gaulin, C.; Lê, M.; Shum, M.; Fong, D. *Disinfectants and Sanitizers for Use on Food Contact Surfaces*; National Collaborating Centre for Environmental Health: Vancouver, BC, Canada, 2011; pp. 1–15.
73. Bisbiroulas, P.; Psylou, M.; Iliopoulou, I.; Diakogiannis, I.; Berberi, A.; Mastronicolis, S.K. Adaptational changes in cellular phospholipids and fatty acid composition of the food pathogen *Listeria monocytogenes* as a stress response to disinfectant sanitizer benzalkonium chloride. *Lett. Appl. Microbiol.* **2011**, *52*, 275–280. [CrossRef] [PubMed]
74. Møretrø, T.; Schirmer, B.C.T.; Heir, E.; Fagerlund, A.; Hjemli, P.; Langsrud, S. Tolerance to quaternary ammonium compound disinfectants may enhance growth of *Listeria monocytogenes* in the food industry. *Int. J. Food Microbiol.* **2017**, *241*, 215–224. [CrossRef] [PubMed]
75. Gerba, C.P. Quaternary Ammonium Biocides: Efficacy in Application. *Appl. Environ. Microbiol.* **2015**, *81*, 464–469. [CrossRef] [PubMed]
76. Martinez-Suarez, J.V.; Ortiz, S.; López-Alonso, V. Potential impact of the resistance to quaternary ammonium disinfectants on the persistence of *Listeria monocytogenes* in food processing environments. *Front. Microbiol.* **2016**, *7*, 1–8. [CrossRef] [PubMed]
77. Liu, X.; Marrakchi, M.; Jahne, M.; Rogers, S.; Andreescu, S. Real-time investigation of antibiotics-induced oxidative stress and superoxide release in bacteria using an electrochemical biosensor. *Free Radic. Biol. Med.* **2016**, *91*, 25–33. [CrossRef] [PubMed]
78. Shapiro, R.S. Antimicrobial-Induced DNA Damage and Genomic Instability in Microbial Pathogens. *PLoS Pathog.* **2015**, *11*, e1004678. [CrossRef] [PubMed]
79. Dutta, V.; Elhanafi, D.; Kathariou, S. Conservation and distribution of the benzalkonium chloride resistance cassette bcrABC in *Listeria monocytogenes*. *Appl. Environ. Microbiol.* **2013**, *79*, 6067–6074. [CrossRef] [PubMed]
80. Aase, B.; Sundheim, G.; Langsrud, S.; Rørvik, L.M. Occurrence of and a possible mechanism for resistance to a quaternary ammonium compound in *Listeria monocytogenes*. *Int. J. Food Microbiol.* **2000**, *62*, 57–63. [CrossRef]
81. Ebner, R.; Stephan, R.; Althaus, D.; Brisse, S.; Maury, M.; Tasara, T. Phenotypic and genotypic characteristics of *Listeria monocytogenes* strains isolated during 2011–2014 from different food matrices in Switzerland. *Food Control* **2015**, *57*, 321–326. [CrossRef]

82. Kovacevic, J.; Ziegler, J.; Walecka-Zacharska, E.; Reimer, A.; Kitts, D.D.; Gilmour, M.W. Tolerance of *Listeria monocytogenes* to quaternary ammonium sanitizers is mediated by a novel efflux pump encoded by emrE. *Appl. Environ. Microbiol.* **2015**, *82*, 939–953. [CrossRef] [PubMed]

83. Müller, A.; Rychli, K.; Muhterem-Uyar, M.; Zaiser, A.; Stessl, B.; Guinane, C.M.; Cotter, P.D.; Wagner, M.; Schmitz-Esser, S. Tn6188—A Novel Transposon in *Listeria monocytogenes* Responsible for Tolerance to Benzalkonium Chloride. *PLoS ONE* **2013**, *8*, e768635. [CrossRef] [PubMed]

84. Müller, A.; Rychli, K.; Zaiser, A.; Wieser, C.; Wagner, M.; Schmitz-Esser, S. The *Listeria monocytogenes* transposon Tn6188 provides increased tolerance to various quaternary ammonium compounds and ethidium bromide. *FEMS Microbiol. Lett.* **2014**, *361*, 166–173. [CrossRef] [PubMed]

85. Yoon, Y.; Lee, H.; Lee, S.; Kim, S.; Choi, K.H. Membrane fluidity-related adaptive response mechanisms of foodborne bacterial pathogens under environmental stresses. *Food Res. Int.* **2015**, *72*, 25–36. [CrossRef]

86. Dubois-Brissonnet, F.; Trotier, E.; Briandet, R. The Biofilm Lifestyle Involves an Increase in Bacterial Membrane Saturated Fatty Acids. *Front. Microbiol.* **2016**, *7*, 1–8. [CrossRef] [PubMed]

87. Miladi, H.; Ammar, E.; Ben Slama, R.; Sakly, N.; Bakhrouf, A. Influence of freezing stress on morphological alteration and biofilm formation by *Listeria monocytogenes*: Relationship with cell surface hydrophobicity and membrane fluidity. *Arch. Microbiol.* **2013**, *195*, 705–715. [CrossRef] [PubMed]

88. Vaid, R.; Linton, R.H.; Morgan, M.T. Comparison of inactivation of *Listeria monocytogenes* within a biofilm matrix using chlorine dioxide gas, aqueous chlorine dioxide and sodium hypochlorite treatments. *Food Microbiol.* **2010**, *27*, 979–984. [CrossRef] [PubMed]

89. Wei, C.-I.; Cook, D.L.; Kirk, J.R. Use of chlorine compounds in the food industry. *Food Technol.* **1985**, *39*, 107–115.

90. Valderrama, W.B.; Mills, E.W.; Cutter, C.N. Efficacy of chlorine dioxide against *Listeria monocytogenes* in brine chilling solutions. *J. Food Prot.* **2009**, *72*, 2272–2277. [CrossRef] [PubMed]

91. El-Kest, S.E.; Marth, E.H. Inactivation of Listeria Monocytogenes by Chlorine. *J. Food Prot.* **1988**, *51*, 520–524. [CrossRef]

92. Brackett, R.E. Antimicrobial effect of chlorine on *Listeria monocytogenes*. *J. Food Prot.* **1987**, *50*, 999–1003. [CrossRef]

93. Taormina, P.J.; Beuchat, L.R. Survival and heat resistance of *Listeria monocytogenes* after exposure to alkali and chlorine. *Appl. Environ. Microbiol.* **2001**, *67*, 2555–2563. [CrossRef] [PubMed]

94. Lundén, J.; Autio, T.; Markkula, A.; Hellström, S.; Korkeala, H. Adaptive and cross-adaptive responses of persistent and non-persistent *Listeria monocytogenes* strains to disinfectants. *Int. J. Food Microbiol.* **2003**, *82*, 265–272. [CrossRef]

95. Aarnisalo, K.; Salo, S.; Miettinen, H.; Suihko, M.-L.; Wirtanen, G.; Autio, T.; Lundén, J.; Korkeala, H.; Sjoberg, A.-M. Bacterial efficiencies of commercial disinfectants against *Listeria monocytogenes* on surfaces. *J. Food Saf.* **2000**, *20*, 237–250. [CrossRef]

96. Bremer, P.J.; Monk, I.; Butler, R. Inactivation of *Listeria monocytogenes*/*Flavobacterium* spp. biofilms using chlorine: Impact of substrate, pH, time and concentration. *Lett. Appl. Microbiol.* **2002**, *35*, 321–325. [CrossRef] [PubMed]

97. Pan, Y.; Breidt, F.; Kathariou, S. Resistance of *Listeria monocytogenes* biofilms to sanitizing agents in a simulated food processing environment. *Appl. Environ. Microbiol.* **2006**, *72*, 7711–7717. [CrossRef] [PubMed]

98. Folsom, J.P.; Frank, J.F. Chlorine Resistance of *Listeria monocytogenes* Biofilms and Relationship to Subtype, Cell Density, and Planktonic Cell Chlorine Resistance. *J. Food Prot.* **2006**, *69*, 1292–1296. [CrossRef] [PubMed]

99. Norwood, D.E.; Gilmour, A. The growth and resistance to sodium hypochlorite of *Listeria monocytogenes* in a steady-state multispecies biofilm. *J. Appl. Microbiol.* **2000**, *88*, 512–520. [CrossRef] [PubMed]

100. Denyer, S.P.; Stewart, G.S.A.B. Mechanisms of action of disinfectants. *Int. Biodeterior. Biodegrad.* **1998**, *41*, 261–268. [CrossRef]

101. Maillard, J.Y. Bacterial target sites for biocide action. *J. Appl. Microbiol. Symp. Suppl.* **2002**, *92*, 16S–27S. [CrossRef]

102. Cotter, P.D.; Hill, C. Surviving the Acid Test: Responses of Gram-Positive Bacteria to Low pH. *Microbiol. Mol. Biol. Rev.* **2003**, *67*, 429–453. [CrossRef] [PubMed]

103. Ryan, S.; Hill, C.; Gahan, C.G.M. Acid Stress Responses in *Listeria monocytogenes*. In *Advances in Applied Microbiology*; Laskin, A.I., Sariaslani, S., Gadd, G.M., Eds.; Elsevier Academic Press: San Diego, CA, USA, 2008; pp. 67–91.

104. Phan-Thanh, L.; Mahouin, F.; Aligé, S. Acid responses of *Listeria monocytogenes*. *Int. J. Food Microbiol.* **2000**, *55*, 121–126. [CrossRef]

105. Koutsoumanis, K.P.; Kendall, P.A.; Sofos, J.N. Effect of Food Processing-Related Stresses on Acid Tolerance of *Listeria monocytogenes*. *Appl. Environ. Microbiol.* **2003**, *69*, 7514–7516. [CrossRef] [PubMed]

106. Phan-Thanh, L.; Mahouin, F. A proteomic approach to study the acid response in *Listeria monocytogenes*. *Electrophoresis* **1999**, *20*, 2214–2224. [CrossRef]

107. Cotter, P.D.; O'Reilly, K.; Hill, C. Role of the glutamate decarboxylase acid resistance system in the survival of *Listeria monocytogenes* LO28 in low pH foods. *J. Food Prot.* **2001**, *64*, 1362–1368. [CrossRef] [PubMed]

108. Cotter, P.D.; Gahan, C.G.M.; Hill, C. Analysis of the role of the *Listeria monocytogenes* F0F1-ATPase operon in the acid tolerance response. *Int. J. Food Microbiol.* **2000**, *60*, 137–146. [CrossRef]

109. Giotis, E.S.; McDowell, D.A.; Blair, I.S.; Wilkinson, B.J. Role of branched-chain fatty acids in pH stress tolerance in *Listeria monocytogenes*. *Appl. Environ. Microbiol.* **2007**, *73*, 997–1001. [CrossRef] [PubMed]

110. Zhang, Y.; Carpenter, C.E.; Broadbent, J.R.; Luo, X. Influence of habituation to inorganic and organic acid conditions on the cytoplasmic membrane composition of *Listeria monocytogenes*. *Food Control* **2015**, *55*, 49–53. [CrossRef]

111. Saá Ibusquiza, P.; Herrera, J.J.R.; Cabo, M.L. Resistance to benzalkonium chloride, peracetic acid and nisin during formation of mature biofilms by *Listeria monocytogenes*. *Food Microbiol.* **2011**, *28*, 418–425. [CrossRef] [PubMed]

112. Lee, S.; Cappato, L.; Corassin, C.; Cruz, A.; Oliveira, C. Effect of peracetic acid on biofilms formed by Staphylococcus aureus and *Listeria monocytogenes* isolated from dairy plants. *J. Dairy Sci.* **2016**, *99*, 2384–2390. [CrossRef] [PubMed]

113. Hwa, S.; Lee, I.; Barancelli, G.V.; Corassin, C.H.; Rosim, R.E.; Sengling, C.F.; Coppa, C.; Fernandes De Oliveira, C.A. Effect of peracetic acid on biofilms formed by *Listeria monocytogenes* strains isolated from a Brazilian cheese processing plant. *Braz. J. Pharm. Sci.* **2017**, *53*, e00071. [CrossRef]

114. Chorianopoulos, N.; Giaouris, E.; Grigoraki, I.; Skandamis, P.; Nychas, G.J. Effect of acid tolerance response (ATR) on attachment of *Listeria monocytogenes* Scott A to stainless steel under extended exposure to acid or/and salt stress and resistance of sessile cells to subsequent strong acid challenge. *Int. J. Food Microbiol.* **2011**, *145*, 400–406. [CrossRef] [PubMed]

115. Cataldo, G.; Conte, M.P.; Chiarini, F.; Seganti, L.; Ammendolia, M.G.; Superti, F.; Longhi, C. Acid adaptation and survival of *Listeria monocytogenes* in Italian-style soft cheeses. *J. Appl. Microbiol.* **2007**, *103*, 185–193. [CrossRef] [PubMed]

116. Stopforth, J.D.; Samelis, J.; Sofos, J.N.; Kendall, P.A.; Smith, G.C. Biofilm formation by acid-adapted and nonadapted *Listeria monocytogenes* in fresh beef decontamination washings and its subsequent inactivation with sanitizers. *J. Food Prot.* **2002**, *65*, 1717–1727. [CrossRef] [PubMed]

117. Van der Veen, S.; Abee, T. Mixed species biofilms of *Listeria monocytogenes* and Lactobacillus plantarum show enhanced resistance to benzalkonium chloride and peracetic acid. *Int. J. Food Microbiol.* **2011**, *144*, 421–431. [CrossRef] [PubMed]

118. Liu, W.; Røder, H.L.; Madsen, J.S.; Bjarnsholt, T.; Sørensen, S.J.; Burmølle, M. Interspecific bacterial interactions are reflected in multispecies biofilm spatial organization. *Front. Microbiol.* **2016**, *7*, 1366. [CrossRef] [PubMed]

119. Giaouris, E.; Heir, E.; Desvaux, M.; Hébraud, M.; Møretrø, T.; Langsrud, S.; Doulgeraki, A.; Nychas, G.-J.; Kačániová, M.; Czaczyk, K.; et al. Intra- and inter-species interactions within biofilms of important foodborne bacterial pathogens. *Front. Microbiol.* **2015**, *6*, 841. [CrossRef] [PubMed]

120. Kalmokoff, M.L.; Austin, J.W.; Wan, X.D.; Sanders, G.; Banerjee, S.; Farber, J.M. Adsorption, attachment and biofilm formation among isolates of listeria monocytogenes using model conditions. *J. Appl. Microbiol.* **2001**, *91*, 725–734. [CrossRef] [PubMed]

121. Carpentier, B.; Cerf, O. Review—Persistence of *Listeria monocytogenes* in food industry equipment and premises. *Int. J. Food Microbiol.* **2011**, *145*, 1–8. [CrossRef] [PubMed]

122. Norwood, D.E.; Gilmour, A. The differential adherence capabilities of two *Listeria monocytogenes* strains in monoculture and multispecies biofilms as a function of temperature. *Lett. Appl. Microbiol.* **2001**, *33*, 320–324. [CrossRef] [PubMed]

123. Rodríguez-López, P.; Saá-Ibusquiza, P.; Mosquera-Fernández, M.; López-Cabo, M. *Listeria monocytogenes*—carrying consortia in food industry. Composition, subtyping and numerical characterisation of mono-species biofilm dynamics on stainless steel. *Int. J. Food Microbiol.* **2015**, *206*, 84–95. [CrossRef] [PubMed]

124. Bremer, P.J.; Monk, I.; Osborne, C.M. Survival of *Listeria monocytogenes* attached to stainless steel surfaces in the presence or absence of *Flavobacterium* spp. *J. Food Prot.* **2001**, *64*, 1369–1376. [CrossRef] [PubMed]

125. Carpentier, B.; Chassaing, D. Interactions in biofilms between *Listeria monocytogenes* and resident microorganisms from food industry premises. *Int. J. Food Microbiol.* **2004**, *97*, 111–122. [CrossRef] [PubMed]

126. Nilsson, L.; Hansen, T.B.; Garrido, P.; Buchrieser, C.; Glaser, P.; Knøchel, S.; Gram, L.; Gravesen, A. Growth inhibition of *Listeria monocytogenes* by a nonbacteriocinogenic Carnobacterium piscicola. *J. Appl. Microbiol.* **2004**. [CrossRef]

127. Bas, S.; Kramer, M.; Stopar, D. Biofilm surface density determines biocide effectiveness. *Front. Microbiol.* **2017**, *8*, 2443. [CrossRef] [PubMed]

128. Jahid, I.K.; Ha, S.-D. The Paradox of Mixed-Species Biofilms in the Context of Food Safety. *Compr. Rev. Food Sci. Food Saf.* **2014**, *13*, 990–1011. [CrossRef]

129. Midelet, G.; Kobilinsky, A.; Carpentier, B. Construction and analysis of fractional multifactorial designs to study attachment strength and transfer of *Listeria monocytogenes* from pure or mixed biofilms after contact with a solid model food. *Appl. Environ. Microbiol.* **2006**, *72*, 2313–2321. [CrossRef] [PubMed]

130. Sasahara, K.C.; Zottola, E.A. Biofilm formation by *Listeria monocytogenes* utilizes a primary colonizing microorganism in flowing systems. *J. Food Prot.* **1993**, *56*, 1022–1028. [CrossRef]

131. Puga, C.H.; SanJose, C.; Orgaz, B. Spatial distribution of *Listeria monocytogenes* and Pseudomonas fluorescens in mixed biofilms. In *Listeria monocytogenes: Food Sources, Prevalence and Management Strategies*; Hambrick, E.C., Ed.; Nova Publishers: New York, NY, USA, 2014; pp. 115–132. ISBN 9781631170546.

132. Saá Ibusquiza, P.; Herrera, J.J.R.; Vázquez-Sánchez, D.; Cabo, M.L. Adherence kinetics, resistance to benzalkonium chloride and microscopic analysis of mixed biofilms formed by *Listeria monocytogenes* and Pseudomonas putida. *Food Control* **2012**, *25*, 202–210. [CrossRef]

133. Papaioannou, E.; Giaouris, E.D.; Berillis, P.; Boziaris, I.S. Dynamics of biofilm formation by *Listeria monocytogenes* on stainless steel under mono-species and mixed-culture simulated fish processing conditions and chemical disinfection challenges. *Int. J. Food Microbiol.* **2018**, *267*, 9–19. [CrossRef] [PubMed]

134. Langsrud, S.; Sidhu, M.S.; Heir, E.; Holck, A.L. Bacterial disinfectant resistance—A challenge for the food industry. *Int. Biodeterior. Biodegrad.* **2003**, *51*, 283–290. [CrossRef]

135. Ortega Morente, E.; Fernández-Fuentes, M.A.; Grande Burgos, M.J.; Abriouel, H.; Pérez Pulido, R.; Gálvez, A. Biocide tolerance in bacteria. *Int. J. Food Microbiol.* **2013**, *162*, 13–25. [CrossRef] [PubMed]

136. Frentzel, H.; Menrath, A.; Tomuzia, K.; Braeunig, J.; Appel, B. Decontamination of High-risk Animal and Zoonotic Pathogens. *Biosecur. Bioterror. Biodef. Strateg. Pract. Sci.* **2013**, *11*, S102–S114. [CrossRef] [PubMed]

137. Simões, M.; Simões, L.C.; Vieira, M.J. A review of current and emergent biofilm control strategies. *LWT Food Sci. Technol.* **2010**, *43*, 573–583. [CrossRef]

138. Coughlan, L.M.; Cotter, P.D.; Hill, C.; Alvarez-ordóñez, A. New weapons to fight old enemies: Novel strategies for the (bio)control of bacterial biofilms in the food industry. *Front. Microbiol.* **2016**, *7*, 1–21. [CrossRef] [PubMed]

139. Oloketuyi, S.F.; Khan, F. Inhibition strategies of *Listeria monocytogenes* biofilms—Current knowledge and future outlooks. *J. Basic Microbiol.* **2017**, *57*, 728–743. [CrossRef] [PubMed]

140. Møretrø, T.; Langsrud, S. Residential Bacteria on Surfaces in the Food Industry and Their Implications for Food Safety and Quality. *Compr. Rev. Food Sci. Food Saf.* **2017**, *16*, 1022–1041. [CrossRef]

141. Fox, E.M.; Solomon, K.; Moore, J.E.; Wall, P.G.; Fanning, S. Phylogenetic profiles of in-house microflora in drains at a food production facility: Comparison and biocontrol implications of listeria-positive and -negative bacterial populations. *Appl. Environ. Microbiol.* **2014**, *80*, 3369–3374. [CrossRef] [PubMed]

142. O'Sullivan, L.; Ross, R.P.; Hill, C. Potential of bacteriocin-producing lactic acid bacteria for improvements in food safety and quality. *Biochimie* **2002**, *84*, 593–604. [CrossRef]

143. Castellano, P.; Pérez Ibarreche, M.; Blanco Massani, M.; Fontana, C.; Vignolo, G. Strategies for Pathogen Biocontrol Using Lactic Acid Bacteria and Their Metabolites: A Focus on Meat Ecosystems and Industrial Environments. *Microorganisms* **2017**, *5*, 38. [CrossRef] [PubMed]

144. Fontana, C.; Cocconcelli, P.S.; Vignolo, G.; Saavedra, L. Occurrence of antilisterial structural bacteriocins genes in meat borne lactic acid bacteria. *Food Control* **2015**, *47*, 53–59. [CrossRef]

145. Bower, C.K.; McGuire, J.; Daeschel, M.A. Suppression of *Listeria monocytogenes* colonization following adsorption of nisin onto silica surfaces. *Appl. Environ. Microbiol.* **1995**, *61*, 992–997. [PubMed]
146. Minei, C.C.; Gomes, B.C.; Ratti, R.P.; D'Angelis, C.E.M.; De Martinis, E.C.P. Influence of peroxyacetic acid and nisin and coculture with Enterococcus faecium on *Listeria monocytogenes* biofilm formation. *J. Food Prot.* **2008**, *71*, 634–638. [CrossRef] [PubMed]
147. Bolocan, A.S.; Pennone, V.; O'Connor, P.M.; Coffey, A.; Nicolau, A.I.; McAuliffe, O.; Jordan, K. Inhibition of *Listeria monocytogenes* biofilms by bacteriocin-producing bacteria isolated from mushroom substrate. *J. Appl. Microbiol.* **2017**, *122*, 279–293. [CrossRef] [PubMed]
148. Al-Seraih, A.; Belguesmia, Y.; Baah, J.; Szunerits, S.; Boukherroub, R.; Drider, D. Enterocin B3A–B3B produced by LAB collected from infant faeces: Potential utilization in the food industry for *Listeria monocytogenes* biofilm management. *Antonie van Leeuwenhoek* **2017**, *110*, 205–219. [CrossRef] [PubMed]
149. Caballero Gómez, N.; Abriouel, H.; Grande, M.J.; Pérez Pulido, R.; Gálvez, A. Effect of enterocin AS-48 in combination with biocides on planktonic and sessile *Listeria monocytogenes*. *Food Microbiol.* **2012**, *30*, 51–58. [CrossRef] [PubMed]
150. Winkelströter, L.K.; Gomes, B.C.; Thomaz, M.R.S.; Souza, V.M.; De Martinis, E.C.P. Lactobacillus sakei 1 and its bacteriocin influence adhesion of *Listeria monocytogenes* on stainless steel surface. *Food Control* **2011**, *22*, 1404–1407. [CrossRef]
151. Leriche, V.; Chassaing, D.; Carpentier, B. Behaviour of *L. monocytogenes* in an artificially made biofilm of a nisin-producing strain of *Lactococcus lactis*. *Int. J. Food Microbiol.* **1999**, *51*, 169–182. [CrossRef]
152. Habimana, O.; Meyrand, M.; Meylheuc, T.; Kulakauskas, S.; Briandet, R. Genetic features of resident biofilms determine attachment of *Listeria monocytogenes*. *Appl. Environ. Microbiol.* **2009**, *75*, 7814–7821. [CrossRef] [PubMed]
153. Zhao, T.; Doyle, M.P.; Zhao, P. Control of *Listeria monocytogenes* in a biofilm by competitive-exclusion microorganisms. *Appl. Environ. Microbiol.* **2004**, *70*, 3996–4003. [CrossRef] [PubMed]
154. Zhao, T.; Podtburg, T.C.; Zhao, P.; Chen, D.; Baker, D.A.; Cords, B.; Doyle, M.P. Reduction by competitive bacteria of *Listeria monocytogenes* in biofilms and Listeria bacteria in floor drains in a ready-to-eat poultry processing plant. *J. Food Prot.* **2013**, *76*, 601–607. [CrossRef] [PubMed]
155. García-Almendárez, B.E.; Cann, I.K.O.; Martin, S.E.; Guerrero-Legarreta, I.; Regalado, C. Effect of Lactococcus lactis UQ2 and its bacteriocin on *Listeria monocytogenes* biofilms. *Food Control* **2008**, *19*, 670–680. [CrossRef]
156. Guerrieri, E.; de Niederhäusern, S.; Messi, P.; Sabia, C.; Iseppi, R.; Anacarso, I.; Bondi, M. Use of lactic acid bacteria (LAB) biofilms for the control of *Listeria monocytogenes* in a small-scale model. *Food Control* **2009**, *20*, 861–865. [CrossRef]
157. Ratti, R.P.; Gomes, B.C.; Martinez, R.C.R.; Souza, V.M.; Martinis, E.C.P. De Elongated cells of *Listeria monocytogenes* in biofilms in the presence of sucrose and bacteriocin-producing Leuconostoc mesenteroides A11. *Ciênc. Tecnol. Aliment.* **2010**, *30*, 1011–1016. [CrossRef]
158. Pérez-Ibarreche, M.; Castellano, P.; Leclercq, A.; Vignolo, G. Control of *Listeria monocytogenes* biofilms on industrial surfaces by the bacteriocin-producing Lactobacillus sakei CRL1862. *FEMS Microbiol. Lett.* **2016**, *363*, 1–6. [CrossRef] [PubMed]
159. Ndahetuye, J.B.; Koo, O.K.; O'bryan, C.A.; Ricke, S.C.; Crandall, P.G. Role of Lactic Acid Bacteria as a Biosanitizer to Prevent Attachment of *Listeria monocytogenes* F6900 on Deli Slicer Contact Surfaces. *J. Food Prot.* **2012**, *75*, 1429–1436. [CrossRef] [PubMed]
160. Woo, J.; Ahn, J. Probiotic-mediated competition, exclusion and displacement in biofilm formation by food-borne pathogens. *Lett. Appl. Microbiol.* **2013**, *56*, 307–313. [CrossRef] [PubMed]
161. Winkelströter, L.K.; Tulini, F.L.; De Martinis, E.C.P. Identification of the bacteriocin produced by cheese isolate Lactobacillus paraplantarum FT259 and its potential influence on *Listeria monocytogenes* biofilm formation. *LWT Food Sci. Technol.* **2015**, *64*, 586–592. [CrossRef]
162. Kanmani, P.; Satish Kumar, R.; Yuvaraj, N.; Paari, K.A.; Pattukumar, V.; Arul, V. Probiotics and Its Functionally Valuable Products—A Review. *Crit. Rev. Food Sci. Nutr.* **2013**, *53*, 641–658. [CrossRef] [PubMed]
163. Guillier, L.; Stahl, V.; Hezard, B.; Notz, E.; Briandet, R. Modelling the competitive growth between *Listeria monocytogenes* and biofilm microflora of smear cheese wooden shelves. *Int. J. Food Microbiol.* **2008**, *128*, 51–57. [CrossRef] [PubMed]

164. Habimana, O.; Guillier, L.; Kulakauskas, S.; Briandet, R. Spatial competition with Lactococcus lactis in mixed-species continuous-flow biofilms inhibits *Listeria monocytogenes* growth. *Biofouling* **2011**, *27*, 1065–1072. [CrossRef] [PubMed]

165. Arevalos-Sánchez, M.; Regalado, C.; Martin, S.E.; Domínguez-Domínguez, J.; García-Almendárez, B.E. Effect of neutral electrolyzed water and nisin on *Listeria monocytogenes* biofilms, and on listeriolysin O activity. *Food Control* **2012**, *24*, 116–122. [CrossRef]

166. Henriques, A.R.; Fraqueza, M.J. Biofilm-forming ability and biocide susceptibility of *Listeria monocytogenes* strains isolated from the ready-to-eat meat-based food products food chain. *LWT Food Sci. Technol.* **2017**, *81*, 180–187. [CrossRef]

167. Camargo, A.C.; de Paula, O.A.L.; Todorov, S.D.; Nero, L.A. In Vitro Evaluation of Bacteriocins Activity Against *Listeria monocytogenes* Biofilm Formation. *Appl. Biochem. Biotechnol.* **2016**, *178*, 1239–1251. [CrossRef] [PubMed]

168. Ming, X.; Daeschel, M.A. Nisin Resistance of Foodborne Bacteria and the Specific Resistance Responses of *Listeria monocytogenes* Scott A. *J. Food Prot.* **1993**, *56*, 944–948. [CrossRef]

169. Crandall, A.D.; Montville, T.J. Nisin Resistance in *Listeria monocytogenes* ATCC 700302 Is a Complex Phenotype. *Appl. Environ. Microbiol.* **1998**, *64*, 231–237. [PubMed]

170. Zhao, T.; Podtburg, T.C.; Zhao, P.; Schmidt, B.E.; Baker, D.A.; Cords, B.; Doyle, M.P. Control of *Listeria* spp. by competitive-exclusion bacteria in floor drains of a poultry processing plant. *Appl. Environ. Microbiol.* **2006**, *72*, 3314–3320. [CrossRef] [PubMed]

171. Schöbitz, R.; González, C.; Villarreal, K.; Horzella, M.; Nahuelquín, Y.; Fuentes, R. A biocontroller to eliminate *Listeria monocytogenes* from the food processing environment. *Food Control* **2014**, *36*, 217–223. [CrossRef]

172. Chi-Zhang, Y.; Yam, K.L.; Chikindas, M.L. Effective control of *Listeria monocytogenes* by combination of nisin formulated and slowly released into a broth system. *Int. J. Food Microbiol.* **2004**, *90*, 15–22. [CrossRef]

173. Gravesen, A.; Kallipolitis, B.; Holmstrøm, K.; Høiby, P.E.; Ramnath, M.; Knøchel, S. pbp2229-Mediated Nisin Resistance Mechanism in *Listeria monocytogenes* Confers Cross-Protection to Class IIa Bacteriocins and Affects Virulence Gene Expression. *Appl. Environ. Microbiol.* **2004**, *70*, 1669–1679. [CrossRef] [PubMed]

174. Naghmouchi, K.; Kheadr, E.; Lacroix, C.; Fliss, I. Class I/Class IIa bacteriocin cross-resistance phenomenon in *Listeria monocytogenes*. *Food Microbiol.* **2007**, *24*, 718–727. [CrossRef] [PubMed]

175. Belguesmia, Y.; Madi, A.; Sperandio, D.; Merieau, A.; Feuilloley, M.; Prévost, H.; Drider, D.; Connil, N. Growing insights into the safety of bacteriocins: The case of enterocin S37. *Res. Microbiol.* **2011**, *162*, 159–163. [CrossRef] [PubMed]

176. Marder, E.P.; Griffin, P.M.; Cieslak, P.R.; Dunn, J.; Hurd, S.; Jervis, R.; Lathrop, S.; Muse, A.; Ryan, P.; Smith, K.; et al. Preliminary Incidence and Trends of Infection with Pathogens Transmitted Commonly through Food—Foodborne Diseases Active Surveillance Network, 10 U.S. Sites, 2006–2017. *Morb. Mortal. Wkly. Rep.* **2018**, *64*, 324–328. [CrossRef] [PubMed]

177. Nguyen, U.T.; Burrows, L.L. DNase I and proteinase K impair *Listeria monocytogenes* biofilm formation and induce dispersal of pre-existing biofilms. *Int. J. Food Microbiol.* **2014**, *187*, 26–32. [CrossRef] [PubMed]

178. Gray, J.A.; Chandry, P.S.; Kaur, M.; Kocharunchitt, C.; Bowman, J.P.; Fox, E.M. Novel Biocontrol Methods for *Listeria monocytogenes* Biofilms in Food Production Facilities. *Front. Microbiol.* **2018**, *9*, 605. [CrossRef] [PubMed]

179. Rodríguez-López, P.; Puga, C.H.; Orgaz, B.; Cabo, M.L. Quantifying the combined effects of pronase and benzalkonium chloride in removing late-stage *Listeria monocytogenes–Escherichia coli* dual-species biofilms. *Biofouling* **2017**, *33*, 690–702. [CrossRef] [PubMed]

MDPI

St. Alban-Anlage 66

4052 Basel

Switzerland

Tel. +41 61 683 77 34

Fax +41 61 302 89 18

www.mdpi.com

Foods Editorial Office

E-mail: foods@mdpi.com

www.mdpi.com/journal/foods